D0603951

infection

Also by Gerald N. Callahan, Ph.D.

❀

*Faith, Madness, and Spontaneous
    Human Combustion*

River Odyssey

infection

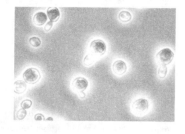

THE

UNINVITED

UNIVERSE

Gerald N. Callahan, Ph.D.

ST. MARTIN'S PRESS
NEW YORK

www.stmartins.com

Design by Kathryn Parise

LIBRARY OF CONGRESS CATALOGING-IN-PUBLICATION DATA

Callahan, Gerald N., 1946–
    Infection : the uninvited universe / Gerald N. Callahan.
      p. cm.
    Includes bibliographical references (p.).
    ISBN-13: 978-0-312-34846-5
    ISBN-10: 0-312-34846-0

  RC111.C25 2006
  616.9—dc22

                        2006045058

First Edition: December 2006

10 9 8 7 6 5 4 3 2 1

To my wife, Gina, who made this book—along with most everything I am proud of in my life—possible

Contents ╍•

Acknowledgments

Several people have helped me make this book better. I would like to thank Dr. Ron Iverson and Joannah Merriman for careful reading of parts of this manuscript and for their helpful criticism. I would like to thank Ethan Friedman at St. Martin's Press for his thoughtful suggestions and his help with the book's final form. I especially want to thank Jane Dystel and Miriam Goderich at Dystel and Goderich Literary Management for their continuing support, their help with the birth of this book, and their commitment to my work. Finally, I want to thank my wife, Gina Mohr-Callahan, for the many hours she devoted to this book—from helping me form the first few ideas to editing the references at the end. It's a much better book because of her.

Introduction ⁝⁝

My Grandfather's Wooden Leg

Charles Gerald Callahan kicked at the gravel as he strode toward his last train. The small rail yard in Coffeyville, Kansas, was overflowing with freight, and the yard chief called Charlie off his job as a conductor to put him to work as a switchman, moving steel cars from track to track. The chief wanted this mess cleared up by the end of the day. Charlie had his doubts. It was well after dark when Charlie walked up to the last car—a load of hogs headed for Chicago. As Charlie checked the manifest, his only light came from the swinging lantern he held and the yellow lightning bugs.

Moving rail cars wasn't particularly complex, but it was time-consuming—uncoupling the cars, moving the last car onto a siding, moving the train back onto the main track, coupling up the rest of the train. And it did require some thought and careful communication between engineer and switchman. But, sooner than he thought, Charlie and the engineer had the hogs onto a spur and the remaining cars recoupled.

That was the last of it. The train, or what remained of it, was bound for Joplin first and would be gone in a moment. Tired from too many hours of moving steel, Charlie waved his lantern, then set it down. He arched his back in a long stretch, then began to roll himself a cigarette. The engineer reversed the locomotive and started forward. The boxcar in front of Charlie bucked and jerked and began to roll just as Charlie realized he'd forgotten to place the locking pin in the couplings.

He dropped the makings of his cigarette and began trotting alongside the train as it rolled south, grabbing for the pin. The iron bolt was hanging on the opposite side of the couplings, and Charlie had to reach twice before he got hold of it. The train was moving faster now.

Charlie started running and trying to thread the pin into its hole. The tracks flashed past below the train and made it hard for him to focus on the couplings. It took him three tries before he finally dropped the pin into place. Then he exhaled and stepped back.

As he did, his foot lodged between the two sets of rails at a switch.

Charlie lurched sideways, trying to clear the tracks, but his foot stayed wedged between the rails. His knee twisted at an odd angle, and he fell onto his back in the coarse gravel. Then he watched bug-eyed as the two sets of steel wheels chewed through the side of his right foot.

About then, the engineer happened to look back, and he saw Charlie spin out from beneath the train and across the gravel. The engineer slowed, then stopped the train, jumped down, and ran toward Charlie.

Charlie was free from the switch, but a slice of his foot was not. The train had carved cleanly through his leather boot, his sock, his skin, his flesh, and his bones. Charlie had dragged himself ten or twelve feet from the tracks and was lying there, propped on his elbows, gulping huge draughts of the warm night air. His foot, or what was left of it, was pumping little arcs of blood onto the oiled gravel. Finally, the engineer reached Charlie.

"You okay, Charlie?"

"Jesus, Hal. Does it look like I'm okay?"

"Sorry. How bad is it?"

"I don't know. God, there's a lot of blood, and I don't think I can walk on it at all," Charlie gasped.

The engineer picked up Charlie's lantern and moved in for a better look. Half of Charlie's right foot still lay in the steel switch. What remained with Charlie was a bloody mass of leather and skin and gristled bone.

"We better get you to the doctor," he said to Charlie, keeping his voice as calm as he could. Then the engineer screamed for help.

Two yardmen answered the engineer's cry, and the three of them carried Charlie to the station house. There they wrapped Charlie's foot in a rag tied as tightly as they could get it. Then they loaded him into a wagon and took him to the doctor's house.

Along the way, two miracles happened: First, Charlie didn't bleed to death, and second, as Charlie lay in the back of that wagon, something too small to be seen crawled in through the hole in Charlie's foot and changed his life forever.

Clostridium perfringens is a rod-shaped bacterium that causes many diseases. Most important for Charlie, this bacterium causes gas gangrene. *C. perfringens* was probably in the soil where Charlie fell or in among the rotten hay and dirt clods in the wagon where Charlie rode. *C. perfringens* is in the dirt nearly everywhere.

When the doctor sutured up Charlie's foot, he created the perfect oxygen-poor home for *C. perfringens*. Inside Charlie, the bacteria multiplied and began to produce toxins and gas. The two together crimped the arteries carrying blood to Charlie's foot. The last of that foot swelled up, turned black, and began to rot. The smell alone was enough to tell everyone that something more had to be done.

All of this happened in about 1910, long before antibiotics. So the doctor's only choice was to take another piece off of Charlie's foot, sew it up again, and pray. Getting ahead of the bacteria proved a difficult task. In a battle to save Charlie, the doctor sliced at Charlie's leg like he might a hard salami, and piece after piece of Charlie fell into the slop bucket.

Finally, after considerable thought, the doctor took off Charlie's leg to the knee. He figured if he cut off a big enough chunk maybe he could beat the demon that was devouring Charlie's leg. Every day, Charlie watched for the rot to reappear. Every day he prayed the rosary and said a few Our Fathers.

This time, *C. perfringens* threw in the towel. Charlie and Doc Wilson had won, but just barely. Charlie would live and walk, after a fashion. In time, Charlie got back to work, using a wooden limb to pole his way across the gravel yards and climb aboard another iron monster headed for Omaha or Joplin or Muskogee.

Charles Gerald Callahan was my grandfather. If he had died from the infection festering in his foot, I wouldn't be writing this. So his story and my brush with oblivion have always fascinated me. When I was a child, I asked him to tell it to me often. Often as I asked, Charlie obliged. He was a patient man.

But every time I listened to his story, it frightened me a little. Even then, I was in awe of the power of infection. I still am. I am who I am because of what Charlie's infections did and didn't do to him. And Charlie was who he was—amputee, Irish Catholic, father, husband, brother, and I'd-rather-die-than-cross-a-picket-line union man because of bacteria. Charlie's is a story of infection.

But, in the end, every human story is a story of infection. *Infection: The Uninvited Universe* is about the billions of microbes that make us human and the few that make us dead. We are not what we seem. Each of us is a bustling community of living things—bacterial things, viral things, fungal things—working together to stay alive in a deadly world. The balance within that community is critical for health, for illness, for the course of world politics and geography. For better or worse, germs have made us who we are, and germs will choose what we shall become.

I've divided the book into three parts. The first section, "Good Germs," describes how, if we are to become human beings, we must be infected—first by our own mothers and then by everything that surrounds us. Without our vermin, we are nothing—they feed us, they protect us, they direct the development of our gastrointestinal and immune systems, they've directed our evolution, increased the sizes of our brains, and laid down the course of our futures.

The second section of the book, "The Lunatic Fringe," explores what happens when things go wrong, when the minority turns against us and infection becomes disease. Only a very, very few bacteria, viruses, and fungi cause human diseases. But a few is all it takes. And the consequences are horrific—AIDS, malaria, leprosy, hemorrhagic fevers, gangrene, blinded eyes, broken blood vessels, rotten flesh, and bloated limbs. The few, the mad, the agents of infectious diseases.

The last section of the book, "Microbes That Will Change the World," examines a group of these maladjusted microorganisms that are changing or will change our lives—SARS, malaria, dengue fever, West Nile virus, anthrax, bubonic plague, bioterrorism, mad cow disease, HIV, and of course, the one with the greatest potential for human devastation, the slate wiper: influenza. Though we ignore them at times, infectious diseases are at the very heart of this world's economy, politics, past, present, and future. None of us will pass through this vale untouched.

Throughout, to smooth the flow of the narrative, I have not included numbered references to indicate my sources for all of the material here. But for anyone interested in identifying or consulting these resources, endnotes identify all of my sources. These appear in order of their occurrence in the text and are correlated with corresponding pages and subject references.

part one

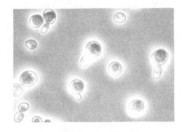

GOOD GERMS

1 ⦂⦂

Infections: Where We Get Them

Henry Perry raises his wineglass to his nose and inhales the rich scent of the claret. Across the rim of the glass he stares into the startling blue eyes of the young woman he bought for tonight. She is dressed in lavender, and her auburn hair falls onto bare shoulders. Henry hasn't seen such pale and perfect flesh in months. Her name is Adrienne, and she smells of musk. She smiles at Henry as his eyes swallow her.

He is enjoying himself. For the past six months he has fought trench foot and Germans in the mud at Soissions. But not tonight. Tonight, Henry's uniform is only for show. His eyes and his nose and his mouth, even his bravery, are for Adrienne only.

Henry places his wineglass next to his linen napkin.

"You are a lovely woman, Adrienne."

"And you are a lovely man," she says to Henry in words wrapped in the syrup of her French.

Henry laughs nervously.

But she is right. Henry is young, blond-haired, and tall. The weeks at war have thinned him, and now he is shaped more like a man than a boy. His uniform, by some monumental accident, fits him perfectly. To-night, Henry's long arms and legs seem just right to Adrienne and the others who have noticed the young American and his escort.

The waiter, in his clipped French, interrupts to ask Henry if he would like more wine. Adrienne translates.

"Tell him I do," Henry says to her. "Tell him I want all the wine he can bring. And then I want you."

Adrienne tells the waiter to bring one more bottle of wine. Beneath the table she adjusts her stockings and straightens her skirt. She reaches across the table for Henry's hand. Oddly, she finds a nugget of eagerness inside her own stomach this evening. *How surprising,* she thinks.

The waiter arrives with another bottle of wine and uncorks it as they both watch. He fills their glasses and leaves.

The lights here are dim, the carpet and curtains thick, the food spiced. Before the war, lovers came here often to sit in dim corners be-neath dark wood and eat from one another's forks.

And that's what those who surround Henry and Adrienne imagine them to be, lovers. But these others, with their own reasons and purposes tonight, are wrong. Henry and Adrienne are not lovers. Theirs is a prac-tical arrangement. Henry and Adrienne have agreed on the cost of love, and it is too high. Both are here for something else, something less ex-pensive, something a soldier on a short furlough can afford. Or so they think.

Everything Henry sees is just as it should be this night. The room, the candlelight, Adrienne, her dress, her ears, her mouth, her neck. The thick red wine and the salty food. Outside, even the darkness seems dressed just for Henry and Adrienne.

The things he cannot see, Henry has decided to ignore. There is so much to leave behind—the Germans, the trenches, pieces of people in the mud. Not tonight.

Tonight, Henry will fill his eyes and thoughts only with Adrienne.

Who could blame him? But in the end, that choice will cost Henry his life.

"Is it time?" he asks.

"Nearly," she says.

Neither of them expects any of this to last beyond tonight. That is what they have planned. That is what they have agreed upon. But there are others here tonight, making plans of their own.

By the time I met Henry, he could no longer speak in complete sentences. He couldn't walk the fifty feet or so across our backyard without jerks and staggers. Sometimes he drooled. These were Adrienne's gifts.

Henry was living, then, at the VA hospital in Salt Lake City, Utah. Every Sunday afternoon, my father would gather him up in our family's Ford station wagon and bring him out to our house in Bountiful for dinner. Henry seemed to like that.

Henry was my mother's brother. She and her mother, who lived with us then, had arranged for Henry to be moved from the VA hospital in Kansas City to the VA in Salt Lake City so that he would be closer to them.

His eyes were still piercingly blue, his hair still blond, though mottled with gray, and he still had the whippet-thin frame of a soldier. But the rest was no longer Henry.

In spite of his peculiarities, or maybe because of them, I enjoyed my uncle. He cursed and spat and wore soiled clothes, all of which I admired. But even I could see that something about Henry was not right.

More than once I asked my mother to explain Henry's peculiarities, but it wasn't until long after his death that she told me the truth. Henry had syphilis. For my mother, that was like a slap in the face. Syphilis was a disease of the poor, the deviant, the unwashed. It was a sickness that fell upon the godless as punishment for their sins.

Syphilis, of course, isn't punishment for anything. It is a disease caused by a bacterium—*Treponema palidum*—a little curlicue of catastrophe known as a spirochete. *T. palidum* moves from person to person during the most intimate of human acts. Wounds, torn tissues, cracked skin are all open doors for syphilis.

That night in Paris, another was sitting at the table with Henry and Adrienne, one no human could see, but one who knew full well that this was not to be a one-night stand. Even Adrienne's eagerness that night was fired by the bacterium. The next morning, as Adrienne stayed behind in her small flat overlooking the Rue Michelet, the spirochete left with Henry.

Fleming's penicillin wouldn't come along for decades, so over the years, *T. palidum* had its way with Henry. First, there was a minor wound, a chancre, not at all painful. And then it went away. Henry was relieved. Later, a rash spread across Henry's palms and the soles of his feet. He assumed it was left over from the trenches of France. Headaches followed. Then the spirochete took Henry's joints—his knuckles, his knees—then it took his eyes, his spine, and his mind. When there was nothing more to take, *T. palidum* took my uncle Henry's life. After all, Henry had dared to take a night off from the war and a moment's pleasure from a beautiful woman in a simple flat one night in Paris.

A microscopic curl of protoplasm, overlooked in the frenzy of lust.

When I finally understood what happened to Henry, I was shocked. Not because Henry had acquired his disease during an act of illicit love, but because of the immensity and the voracity of his infection. The idea that something as tiny and as simple as a bacterium could so unrelentingly and so easily take both a man's mind and his life scared me.

That, of course, changed nothing. A single child's fear was of no consequence to the dominant form of life on planet Earth.

That's simply how things are.

Henry was my uncle. But Henry's story is not unique, not nearly. Bacteria are the most numerous living things on Earth. Everything on the face of this planet, living or dead, has been changed by bacteria— the color of our skies and seas, the air we breathe, the soil beneath our feet, our immune systems, our digestive systems, and each and every human cell. Infection is the way of life. We owe everything we have to bacteria.

The Bacteria That Make Us Human:
Our Normal Flora

I read a science fiction story once in which space travel for humans was possible only when men and women were disassembled into their component cells and stored in vats of salt water. In this way, while rockets accelerated to the enormous speeds needed to reach distant planets during anyone's lifetime, the effects of the massive g-forces were diminished. At the end of such trips, a very complicated computer would suck up each man and each woman from their tubs of brine and reassemble them into human beings.

One young woman, faced with the prospect of dis- and reassembly, expressed considerable concern over the computer's ability to put her back together properly. And worse, she wondered if the computer erred whether she would ever know.

To assure herself of accurate reconstruction, she asked a male friend to go over her body very carefully before and after the space flight. Not surprisingly, he agreed. And in the end, he concluded that the woman was herself once again. This reassured her. But the man's assertions were meaningless. There was no way he could possibly have known if she was or wasn't the same woman.

Each of us is made from billions of little bits of life. In an average person, there are about 1.1×10^{14} of these bits. That's 1.1 with thirteen zeros after it. 1.1×10^{14} is roughly equal to the number of seconds in three million years, twenty thousand times the number of people on Earth, the number of thimblefuls of water in five cubic miles of ocean, or the number of grains in a hundred thousand cubic yards of beach sand. A lot of bits. We call these bits of living things cells—skin cells, red blood cells, white blood cells, liver cells, nerve cells, epithelial cells. Certainly 1.1×10^{14} is more pieces than any man could verify in a lifetime, regardless of how earnestly he might try.

But let's suppose the young man could have checked every single cell in the young woman's body (after all, this was science fiction). When he did, he would have found something startling. After their long space flight and biological metamorphoses, he would have found that the woman's body contained nearly ten times as many bacteria cells as human cells. And if, as he panicked over this discovery, he had checked his own body, he would have found that most of his cells were also bacteria. Imagine their horror. Imagine the setback this would have dealt space travel if word of it had leaked out.

But the bacteria inside both of them had nothing to do with their space flight. Though we don't often notice it, every human is mostly bacteria. In an average human body—even one that has never left Earth—only about 10 percent of the cells are what we call "human" cells. The rest—the vast majority, about 90 percent—are bacteria.

And, actually, not even one of the 10 percent of cells we call human is fully human. Even these cells contain bacteria. That means, as I write these words, by cell number I am less than 10 percent human—and that's stretching it. And as you read these words the same is true of you. Humans are at best only 10 percent human.

How does this happen? How does each one of us come to be so massively infested with these microscopic vermin?

Betrayed at Birth

At the dark moment of fertilization—when, with a final flash of its pearly tail, a sperm penetrates an egg—we are more human than we will ever be again. Sperm, egg, zygote, blastula, embryo, human. And for the next nine months, nestled in the sterile seas of our mothers' wombs, we remain nearly human.

But as birth nears, everything changes. In anticipation of our arrival, soon-to-be mothers begin preparing a special nursery. Inside these women's birth canals, bacteria sprout like weeds. *Lactobacilli*, the same bacteria found in yogurt and buttermilk, divide and spread throughout the passage that we must traverse as we enter this world. Like tiny sausages, about one-one hundredth the thickness of a human hair and as long as one-twenty fifth the thickness of this page, these bacteria rise like fingers to prod us into reality.

Then, on our way into this world these bacteria immediately infect us. Before our mothers have shared so much as a single caress, they have inoculated every one of us, infected us with billions upon billions of squirming, wriggling, living bacteria. Before we draw a first breath, our own mothers have compromised our humanity.

And cruel as that seems, it appears they must. If *Lactobacilli* don't flourish inside a mother's vagina, premature delivery happens more often than in infected mothers. And infants born prematurely or by cesarean section face many challenges that full-term and fully infected infants don't (see Chapter 2). So everything conspires to ensure our immediate infection.

After birth, things get dramatically dirtier. Our first breaths, the arms of the doctor, or midwife, or forest floor, the mother's breasts—all are teeming with microbes. Even in the relative sterility of a hospital delivery room, we roll in the powdered sugar of this world like a warm doughnut fresh from the oil. And as we do, we are quickly covered with layer upon layer of bacteria and fungi and viruses and even a few parasites. Life as a separate entity is over. From this moment on, no one of us ever walks alone.

As a mother nurses her child, she lays the groundwork for further infection. The milk she feeds her child is laced with proteins. Some of these proteins are fertilizers for more infectious microbes, especially *Bifidobacteria*. These bacteria push aside a few of the *Lactobacilli* and attach to the baby's intestines. Together these two bacteria weave a blanket inside the child, a protective blanket.

Children with too few bacteria often develop a disease called oral thrush. *Candida albicans* causes oral thrush. *Candida albicans* is a yeast, the same yeast that causes vaginal yeast infections in women. Without protective bacteria, a child's mouth sprouts thick white crusts of yeast across cheeks and gums and lips. As it spreads, *Candida albicans* digests the human tissues beneath and causes painful destructive ulcers. Without bacteria, life is harder.

The process of our infection appears completely chaotic, but it isn't. *Lactobacilli* and *Bifidobacteria* come first. Then, as the mother withdraws her milk, the *Bifidobacteria* give way to other strains of bacteria, and much later, as the hormones of puberty flood the body, still other bacteria arrive and thrive. Layer upon layer is laid down. Ordered and structured by time and chemistry, we absorb our surroundings. Literally, we become what we eat, or drink, or touch, or breathe. We are what we wear. We become those who caress us and whom we caress in return. From the world that surrounds us, we gather ourselves.

Once a pulp of nearly pure, pale, human cells, we transform into a microbiological metropolis covered with living things whose names read like a Linnaean litany—*Staphylococcus aureus* and *epidermidis; Streptococcus mitis, mutans, viridans, pyogenes,* and *pneumoniae; Trichomonas tenax; Candida albicans; Haemophilus influenzae.*

Our skin sprouts a cornucopia of microorganisms, including at least four different strains of bacteria and several species of fungi. Our eyes gloss over with three or four strains of bacteria. Noses, throats, the upper reaches of our respiratory systems clog with more than six different types of bacteria. Mouths bloom with half a dozen species of bacteria and fungi. Lower urethras fill with more than ten different bacteria, a few fungi, and a parasite or two. But, by far, the greatest numbers of bacteria settle and prosper in our intestines. In places, the bacterial coat in our large intestine is an inch thick. And our feces, by dry weight, are 50 to 60 percent pure bacteria.

While no one notices, and with little or no effort on our part, we become a menagerie, a walking ecosystem, a universe apart. This world is infectious. So whether we like it or not, wherever we go, hosts of others follow.

There is no longer an "I." Now and forever, there is only "we."

And it isn't just the number of microbes that is staggering. The bacterial genome (bacterial DNA) within each of us contains roughly two million distinct bacterial genes.

When scientists sequenced the human genome, they found only about thirty thousand to fifty thousand different human genes strung out along human chromosomes. But the chromosomes our parents gave us do not hold all of the genes inside of us. The bacterial genes we gather far outnumber the genes given to us by our mothers' eggs and our fathers' sperm. Amazingly, nearly 99 percent of the genes inside of human beings aren't human.

Because human cells, like those in our blood and skin and livers and hearts, are about one hundred to one thousand times larger than bacterial cells—by mass and volume—people appear mostly human. But they aren't. And to exclude the bacterial genes from the human genome is not only arbitrary, it's inhuman.

Infection and Individuality

Through processes that begin with our first breaths, we rapidly change from isolated human beings into symbiotic systems with whole communities of living things sharing a single space. What we need we take from the world around us: parents, aunts, uncles, foods, toilets, toothbrushes, tables, automobiles, strollers, sidewalks, trees, dogs, cats, rats, spoons, and spit.

Surprisingly, though, the community of bacteria within one man or one woman is not simply an accidental consequence of birth and geography.

Where we are born is unquestionably important. Infants born in developing countries acquire bacteria that differ considerably from those of infants born in developed parts of the world. Children born in different hospitals may have very different strains of bacteria in their intestines. And breast-fed babies' intestines contain mostly *Bifidobacteria*, while formula-fed babies' intestines have many more potentially dangerous bacteria, such as coliforms, enterococci, and *Bacteroides*. Where we are born and what we eat do make a difference. But our surroundings and our food are not the only factors.

Among humans, identical twins resemble one another most closely. Identical twins, also called monozygotic twins, arise from a single sperm and single egg. These children have the same chromosomes and the same human genes. Usually, identical twins also live in the same home, eat the same food, breathe the same air, and drink the same water. As you might guess, the communities of bacteria inside identical twins are themselves nearly identical.

Beyond monozygotic twins, things change quickly. Even within a single geographic area, the species of bacteria found inside people vary dramatically from person to person. This is true even between married people living in the same home, eating the same food, drinking the same water, and breathing the same air. You and your spouse, partner, brother, or sister house significantly different collections of bacteria. She might have a little more *Staphylococcus aureus* (the cause of toxic shock syndrome) in her nose or vagina than you have. *Candida albicans* might find him a little more attractive than you. *Helicobacter pylori* (part of the cause of peptic ulcers) makes a living for itself in some stomachs and small intestines but not others. *Citrobacter* (which can cause diarrhea and perhaps meningitis) is comfortable with some of us but not others. Our collections of bacteria are unique, perhaps as unique as each one of us. Bacteria, the creepy, crawly, and slimy parts of this world, may be just as important for making human individuals as are brains or genes. Who I am depends on who they are, and vice versa.

The day we arrive in this world, they are waiting for us—our own unique ensemble of bacteria. And we have no choice but to welcome them.

The Bacteria That Stick

Some of the bacteria that colonize us are itinerants, just passing through on their way to an intestine more to their liking. But the majority of the bacteria we attract are in us for the long haul. These bacteria we call normal flora. "Normal," because these species of bacteria are found in "normal" (without obvious disease) people. And "flora," because some bacteria more closely resemble plants than animals.

We acquire our normal floras with no effort whatsoever. We eat, we breathe, we poke our fingers into the soft parts of this world. In the process, we gather billions upon billions of bacteria—as easily as a ship's hull gathers barnacles.

In the Coit Tower in San Francisco, Victor Arnautoff—inspired by Diego Rivera—painted a mural that depicts life on the streets of the city. The painting is done in muted blues and rusts and tans and in Rivera's Depression-era style. In the background, streets stretch off toward oblivion. In the foreground is the pastiche of daily life in the city. A police officer is placing a call to headquarters; a man is reading a newspaper; two uniformed sailors are making their way toward the pier; another man is unloading crates of food next to piles of carrots and lettuce; women are holding hands with their children and moving toward the shops; other men and women move between tasks; a man collects the mail while another has his pocket picked as he checks the time; truckloads of grain arrive nearby; an elevated trolley carries people downtown as trainloads of cargo arrive in the rail yards; factories belch smoke; a fire truck races to a blaze; and women stand on the paved walks and share stories. Although it seems chaotic, it intertwines as artfully as a symphony or a colony of ants. Some provide food, others defense. Some bring news, others clothing. Some carry food. Some are dying. Some worry over war. Some are sleeping. And each relies on the other, each depends on the rest. Move one, and the sense of what is happening changes.

Inside every one of us, this scene is played out day after day after day. Intestines deliver foodstuffs, eyes bring the news, the immune system keeps track of the bad guys, the liver cleans the water, and the red blood cells purify the air. The mind worries. The blood flows.

Like *City Life*, each of those inside of us relies wholly on each of the others. But unlike Arnautoff's mural, inside humans one group stands out from all the rest, outnumbers all the others taken together—

bacteria, the ground stuff of life. Underneath and in between every brick, bacteria thrive. And that mortars it all together.

Our bacteria are not barnacles. They're not just along for the ride. Our bacteria are paying passengers.

My uncle Henry lost his mind because of a fling in France, *Treponema palidum*—a bacterium—and syphilis. One night in 1910, the Denver and Rio Grande Western Railroad sliced open my grandfather's foot and opened the gate for *Clostridium perfringens*—a bacterium. Before it was done with him, that bacterium took all of my grandfather's foot and a sizable chunk of his leg.

Untold thousands of others have lost their lives because of a single breath they took, a fleabite, a drink of water, a wound, and *Mycobacterium tuberculosis* (TB), *Yersinia pestis* (bubonic plague), *Vibrio cholerae* (cholera), or *Clostridium botulinum* (botulism)—all bacteria.

We notice that. Bacteria reach out and slap us in the face then. It's in the papers, the magazines. It's on TV. It's the reason we thought we went after Saddam Hussein. Agents of death and disease, that's what bacteria are. And sometimes, that's true. Some bacteria do make us sick and kill us. But that's only part of the story, only the tiniest part.

Bacteria outnumber humans by a factor of 10^{20} (that's 100,000,000,000,000,000,000). The oceans are full of them, the earth teems with them, we are covered by them. Everything on this planet teems with bacteria. If bacteria were simply malignant agents of disease, none of us would be here today.

It's true that bacteria took my uncle Henry's eyes, then his legs, and finally his life. And that's a terrible story. But bacteria also gave Henry life to begin with—a considerable gift. Later, it was humans who sent my uncle off to war.

2 ⬤⋮⬤

Infections: Why We Need Them—
Spare the Rot and Spoil the Child

This July Tuesday warms me as I walk to my truck. The air smells of pines and the sour seep from the feedlots to the east. A blue jay screams about how great it is to be a blue jay this morning. But today I don't care about blue jays. This morning, I'm watching the ants.

Several small brown hills rise in the seams of my driveway. This happens every summer. Sometime in spring, the small clods push up, and the ants roll them into something that resembles finely ground tobacco. Small holes grant egress and ingress, and a steady stream of wasp-waisted insects carve their own Silk Road, moving leaves, twigs, and bits of bologna from surface to cave.

On Saturdays, I try by various means to obliterate these mounds and to disperse their inhabitants. On Sundays, while I read the paper, both reappear. On Mondays, work takes my mind. On Tuesdays, I insist on taking time to study the ants.

This Tuesday, a regiment of small soldiers has strung out single-file across the driveway toward a rotting plum. A phalanx of black beads has swallowed the plum itself. From under the mound of plum-sucking ants, another line of workers makes its way directly back to the small brown mounds of earth in the cracks between the concrete of my driveway. I open the door to my pickup truck and drop my computer case inside. Then I walk back to where the ants have arrayed themselves. I pick up the plum with its beard of ants, and I move it a yard up the driveway. For a few minutes, everything is chaos. Ants string out from the plum in several directions. The ants leading the march from the hills begin to lash about lazily like the fronds of an underwater fern. Finally the two—the ants stringing from the plum and the ants lashing from the anthills—meet. There appears to be a brief conversation among the lead ants, cursing me, no doubt. Then the regiment realigns itself, locates the plum, and everything continues pretty much as before. Once again, I move the fruit, now more purple pulp than plum, and watch what happens. The earlier scene repeats itself—a few moments of disarray, then single-file order. Twice more I move the plum. Twice more the ants quickly move to solve the problem I have created for them.

Tired of this, and clearly no match for the ants' resolve, I change tactics. I isolate a single ant and, using a leaf, move her six feet from the others. At first, the ant seems totally confused, adrift. She makes some jagged back-and-forth movements, then she begins to circle. As the ant rotates, the circles widen, the size of a peanut, then an almond, then a pecan, then a golf ball, then a tennis ball, then a baseball, then a basketball. Until the orbiting ant at last bumps into a smell. The smell of purpose. Immediately, the lost one aligns herself with the smell and moves toward the plum, her reason for being reestablished.

I choose another one. This ant looks a little like my father—thin in the middle, a little larger behind, and committed. I move her away from the rest. Confusion. Circling. Community and purpose.

I pick one more—a female again, because all worker ants are females—and I carry her a full five yards from all the rest. She hesitates, suspecting a trick. Her legs lift and drop like she has never felt a concrete like this one. A stone at a time, she feels her way. She stops, lifts her head, seems to watch the sun. Then, abruptly, she begins moving in circles that grow even faster than the ones I've seen before. A moment, at most, and she disappears—another period in a long line of commas.

There are no rogue ants here. No one sets off on his or her own, even when forced to face the unknown. None like the dog Buck. No call from the fierce unknown or the pack that roams where no ant has wandered alone. Resistance is futile.

In fact, I realize, there truly is no individual ant here at all. There is only the colony. In a very real sense, there is no such thing as "ant," only "ants." The unit here in my driveway is the colony. No rugged individualists, no Daniel Boone, no Lewis, no Clark, not among ants. An ant without a colony is meaningless.

But that is not the first thing that comes to mind when a person looks at an anthill. We see what appear to be individual ants moving about doing various individual tasks—hauling leaves, tending eggs, rolling grains of sand, gathering bits of plum. Because most of it is buried, and because we're not used to thinking of living things in other ways, we don't see the colony. But it is there, and in one very real sense, it is all that is there.

From the moment we can first hear, we are conditioned to think and then to see that way—to see men and women as individuals. And once we are fully indoctrinated with that concept, we begin to see all animals and plants as individuals—ants included. But they aren't individuals. And neither are we. We are collectives of creatures—some human, some bacterial, some viral, some fungal, some parasitic. Just as there are no ants separate and apart from their colonies, there are no humans separate and apart from the rest of the living world.

The Fearsome Fate of the Uninfected

Without infection, terrible things happen.

Fruit flies, those annoying little things that buzz in and out of your ears while you are trying to lift a peach from your neighbor's tree, normally live a month or two. But uninfected fruit flies—and in particular fruit flies uninfected by bacteria—live, on average, 30 percent shorter lives. For a fruit fly that might mean a loss of ten or twenty days of life. But for you and me, that would translate to about twenty-seven fewer years of life—a noticeable difference, for flies and people.

Flies without bacteria die sooner than infected flies. And the same is true for paramecia, termites, worms, mosquitoes, mice, rats, and, probably, human beings. Infection is as essential for life as a beating heart.

Why is that? What do these vermin do that we cannot live without?

Ultimately, the answer to that question is that all life on this planet relies on bacteria. Life must come to terms with bacteria, because bacteria are everywhere. And then there are fungi and viruses and parasites, the whole microscopic universe that we must contend with. Infection is now and always has been unavoidable. The only beings that have prospered on this planet have done so not because they learned to avoid infection but because they learned to thrive on infection.

Before men and women came along, no creature ever arrived in this world uninfected. It wasn't possible. Every space that any creature might occupy teemed with microscopic life. So, like the music of the spheres, infection had been humming along just beyond the range of human hearing for as long as there had been living things. No animal, no plant ever suffered from uninfection. No disease, no affliction, no syndrome ever arose naturally because an animal failed to become infected. No one noticed, but it was true all the same. It took the curiosity and the technical wizardry of human beings to create an uninfected soul.

Suddenly, we saw what we'd been missing all along.

Scientists created the first germ-free mice more than fifty years ago. It was as though they had opened Pandora's box.

From beneath the lopsided lid of that box, disease after disease rose, and hope followed none of them.

Unexpectedly, the cecum (the first part of the large intestine) in germ-free mice swelled up to several times its normal size. And in a few mice, the cecum became so large that the small intestine wrapped around itself, and the mice died. Just why a lack of bacteria causes cecal swelling isn't clear, but it is clear that these mice died from lack of infection.

After scientists found ways beyond the early afflictions of germ-free mice, illness after illness emerged.

Uninfected animals need more food and water than their infected counterparts. Germ-free rodents need one-third more water than normal rodents. We and other mammals need a lot of water for normal digestion. Much of that water is secreted by our stomachs and small intestines—as much as thirteen quarts per day. To keep us from dehydrating or being forced to drink thirteen quarts of water per day, the large intestine reabsorbs most of the water secreted by the stomach. Apparently, large intestines without bacteria don't absorb water nearly as well as intestines that are lousy with bacteria.

Germ-free mice need 30 percent more calories than normal mice. That's the equivalent of you or I eating four full meals a day instead of three. Thirty percent more fat, 30 percent more sugar, 30 percent more of most everything, just to keep up. Bacteria help us digest high-energy foods like complex sugars. Without microorganisms' help, all of these energy-rich sugars simply pass through us. To make up for that, animals without bacteria must consume a lot more simple sugars and fats.

Interestingly, even though germ-free mice ingest 30 percent more calories than infected mice, the germ-free mice have much lower levels of body fat. And when scientists returned the bacteria to germ-free mice, these mice underwent a 60 percent gain in body fat within fourteen days and often became insulin-resistant. This happens because gut bacteria accelerate absorption of sugars and their conversion to fat. Such a gain would have been a considerable gift for our ancestors who ate rarely and needed to store all the energy they could. It is less of a gift to modern humans who eat early and often. In fact, there is good evidence that the exact composition of our gut bacteria may be a major factor in predisposing some of us to obesity.

Mice without germs don't develop normal intestines. Uninfected intestines don't develop the same cell layers found in normal intestines.

On top of that, the blood supply between the gut and the rest of the body doesn't form properly in these mice. Digested food fails to pass through the walls of these intestines. The mice that must live with these abnormalities often die from malnutrition.

Mice without germs don't develop functional immune systems. Germ-free rabbits fail to develop immune systems altogether. In rabbits, most immune development occurs in the intestines. Rabbit intestines without bacteria do nothing—no immunity, no digestion—nothing.

Germ-free rabbits are often overwhelmed by infections that a normal rabbit would never even notice. The lack of infection that these rabbits suffered as pups renders them defenseless as adults. And if you infect these rabbits as adults, you usually end up with a bloody mass of meat and fur. What they didn't get as pups, you cannot give them as doe or buck. Immunity depends on infection.

Germ-free mice get inflammatory bowel diseases—diseases like Crohn's disease and ulcerative colitis in humans—that normal mice never have. It seems that intestinal bacteria and other microorganisms normally suppress gut inflammation and keep it at an acceptable level.

This suppression allows us humans to maintain a flourishing layer of germs in our intestines without destructive inflammation. From an immunologist's point of view, this is very nearly a miracle. Normally, bacteria light and stoke the coals that burn beneath inflammation (think of strep and staph and influenza). But not in the gut or on the skin. In the gut and on the skin, bacteria somehow strike the chords that signal the fragile harmony between infection, inflammation, and immunity—a piece as intricate as a Bach fugue.

Uninfected mice must be fed vitamins and other nutrients that infected mice make themselves. The bacteria that normally haunt the halls of our intestines give us something we cannot do without.

And mice without germs are much more prone to allergies than normal mice. Under certain conditions, injection of ovalbumin (the major protein in egg whites) causes mice to become allergic to ovalbumin. But if these same mice are first infected with mycobacteria—a particular type of small bacteria—and then injected with ovalbumin, the mice don't develop allergies.

The list goes on. Mice and rabbits without bacteria and fungi and parasites live pitiful lives.

How is it that infection does so much good?

First, the microorganisms themselves create some of the good. Bacteria chew large sugars into smaller ones that we mammals can absorb. Bacteria produce vitamin K as well as many other nutrients that we cannot make on our own. Vitamin K helps blood clot—a fairly significant human endeavor.

Sugars and vitamins. Nutrients, enzymes, and the keys that unlock our potential. Bacteria feed us.

But the greatest good that comes from infection does not come directly from the products of the bacteria themselves. The greatest good comes from the ways the bacteria manipulate us. Most of the effects that bacteria have on animals, including humans, result from bacteria taking control of host genes. That's right. The bacteria that live inside of us have learned to manipulate our genes—bacteria control some of the most intimate and elemental pieces of us.

The effects of intestinal bacteria on absorption, intestinal development, immune development, prevention of infections, the formation of intestinal blood vessels, and water absorption all result from bacterial control of host (human) genes. Bacteria control hundreds of our genes, maybe more. Humans have only about thirty thousand genes total. Bacterial regulation of at least a few hundred of those genes represents a pretty significant concession of command to another species. We may imagine that we are the ones who direct our destinies. But we should not be so quick to discount the billions upon billions of others who live within our borders.

Nor should we fail to recognize the potential of these tireless workers to organize their vast numbers for singular purposes.

The bacteria inside of us do not simply wander about as individual germs trying to think of something useful to do. They get organized.

When individual bacteria attach to intestines or mouths and then multiply, the germs change. As the bacteria spread, they begin to form thick sheets called biofilms. When the biofilms form, the bacteria undergo dramatic genetic changes and produce many products that are not produced by the same bacteria outside of biofilms. Inside of biofilms, bacteria differentiate, become individuals with individual needs and skills. Some of the germs collect into mushroomlike stalks, others secrete thick polymers that surround and protect the biofilm. Then channels open for water and nutrients, and DNA molecules (genes) pass from bacterium to bacterium. In place of a million or a billion individuals, one thing appears, spread across living human tissues—a single entity with a single purpose. A multicellular being swapping stories and genes and proteins and sugars, with itself and with us.

Then a most remarkable event occurs: The bacteria begin to speak among themselves using complex chemicals. This process is called quorum sensing, and it controls many of the functions of bacteria in biofilms.

One striking example of the power of quorum sensing happens in squid. *Euprymna scolopes* is a pink-and-brown mottled squid about the size of your hand. *E. scolopes* thrives near the Hawaiian Islands. Unlike most of its relatives, these squid live in fairly shallow waters near beaches. The squid hunt and feed at night. Because the water where they live is so shallow, a full or partly full moon causes these squid to cast shadows in the sand as they squirt across the shoals. The moving shadows attract predators, predators with a taste for *E. scolopes*.

E. scolopes has solved this problem with talking bacteria. *Vibrio fischeri*, a marine bacterium, grows in high-density biofilms within specialized organs on the underside of *E. scolopes*. And when *V. fischeri* achieves a high enough density within the squid, quorum sensing causes all the bacteria to emit light.

As the glow blossoms, the shadows that swim beneath the squid vanish. Predators become confused and drift off in search of easier prey.

E. scolopes and *V. fischeri*, though they have different names, are a single creature. Neither is anything without the other. And in the end, it is the bacteria and the stories they spread among themselves that save them all.

Inside humans and other mammals, biofilms don't generally light up our lives, but they do help us to survive. We are only beginning to understand all the ways bacteria use biofilms and quorum sensing. Some bacteria use quorum sensing to develop into pathogenic (disease-causing) organisms in humans. Other bacteria use quorum sensing to interact with and control host cells. And still other bacteria use quorum sensing to produce antibiotic factors that protect us from more dangerous infections. Inside of biofilms, bacteria begin to look more like multicellular organisms, more like us. And, at the same time, humans begin to look more like complex mixtures of animal and bacteria—multi- and single-celled beings that speak to one another and choose paths that neither could have walked alone.

The Miracle in the Dirt: The Hygiene Hypothesis

In the 1990s, Erika von Mutius began a study of the children of reunified Germany. Dr. Mutius, a pediatrician, had taken an interest in childhood asthma and allergies and the origins of these diseases. The reunification of East and West Germany gave her the opportunity to compare children who had grown up in relatively clean and healthful environments (West Germany) with children who had grown up under dirtier and less healthful conditions (East Germany). Dr. Mutius had fully expected that the children of East Germany would have more as well as more severe allergies and asthma. What she found was just the opposite. The children who had grown up under the dirtiest conditions had the fewest allergies and asthmas.

Allergies and asthmas are inappropriate immune responses against innocuous environmental elements—like pollen. It isn't clear why some people develop allergies and others don't. Part of the cause appears to be genetic. Dr. Mutius's work with the German children suggested that there was a major environmental factor as well. But the nature of that environmental factor remained cloudy. In later work, Dr. Mutius studied children on farms and compared them to children living in more urban settings. Children who drank fresh milk from the farm and who had regular contact with farm animals were more than ten times less likely to have asthma and more than four times less likely to have hay fever (a common allergy). She and her coworkers concluded, "Long-term and early-life exposure to stables and farm milk induces a strong protective effect against development of asthma [and] hay fever. . . ." No negative effects of farm life were identified.

Dr. Mutius has gone on to show that similar differences in childhood exposures to infections may explain the differences in the rates of asthmas and allergies among children in China.

Very recently Dr. Mutius and other investigators found that childhood exposure to bacteria, particularly a group called gram-negative bacteria, correlates inversely with the frequency of asthmas among school-age children in Austria, Germany, and Switzerland. Specifically, these researchers measured the levels of endotoxin (a product of gram-negative bacteria) in the mattresses of European school-age children. They found that the higher the levels of endotoxin in children's mattresses, the lower the incidence of asthma. The more bacteria the children had been exposed to, the healthier they were. Similarly, investigators in Boston, Massachusetts, collected dust samples from infants' bedroom floors, mattresses, parents' beds, family rooms, and kitchens. The Boston researchers' findings correlated directly with those of Dr. Mutius. New England children exposed to higher levels of endotoxin during the first one to three months of life were much less likely to develop eczema than children in cleaner environments. Eczema, like allergies and asthma, arises when the immune system overreacts. Again, early frequent exposure to bacteria made children healthier and immune systems wiser. Absence of bacteria made things worse.

A group in France further investigated the connection between infection and immunity by giving nursing infants either normal baby formula or fermented baby formula. The fermented milk contained much higher levels of *Bifidobacteria* than the normal infant formula. When these researchers later immunized the children with polio vaccine, the infants fed the fermented formula regularly produced better immune responses against the poliovirus. The additional bacteria somehow helped these infants develop stronger immune systems that produced higher levels of protection against polio.

Clearly, in humans as well as rodents, bacterial infection is essential to the proper development of the immune system. Without it, we fail to protect ourselves, or worse, we turn on ourselves.

Helicobacter pylori is a bacterium that appears commonly among the normal flora of the human gut. The exact function of *H. pylori* is unknown, but there is solid evidence that this bacterium plays a role in the development of peptic ulcers. Nevertheless, a study performed in Britain indicates that when *H. pylori* is intentionally eliminated from people's intestines, there are several negative consequences, including changes that could lead to increased appetite, weight gain, and esophageal reflux. When the stomach or small intestine is compromised—by, say, too much aspirin—*H. pylori* causes ulcers. But without *H. pylori*, appetites may change, weight gain may follow, or the acids of the stomach may move into the esophagus, where they can do incredible damage.

Every year nearly twenty-five hundred children in the United States contract acute lymphoblastic leukemia, or ALL. ALL is the most common cancer of children, accounting for nearly 25 percent of all cancers diagnosed in children. When this cancer develops, a few white-blood-cell precursors take over the bone marrow. Normally, the bone marrow produces all of the many types of cells of the blood—red blood cells, lymphocytes, platelets, monocytes, eosinophils, basophils, neutrophils, and so on. As ALL develops, the bone marrow is forced to produce only a few types of white blood cells, and these white blood cells are rushed into production so quickly that most are completely nonfunctional. Children with ALL become anemic and thrombocytopenic (there are too few platelets in the blood). Platelets play a big role in repairing damaged blood vessels. So these children bruise and hemorrhage very easily. They are unusually susceptible to infections, especially bacterial infections. Without treatment, nearly all would die.

We know very little about the causes of ALL. Genes are clearly part of it. ALL tends to run in families and involves damaged chromosomes (deletions, inversions, etc.). There are also environmental factors that contribute.

Very recently, researchers in the United Kingdom have provided convincing evidence that one of the environmental factors that increase the risk of ALL is lack of early childhood infection. These investigators compared two large groups of children, ages two to fourteen. One group had attended day care, the other had not. The researchers looked for any relationship between how early the children had attended day care and their likelihood of developing ALL. They found that the earlier the children attended day care (less than three months of age), the less likely they were to develop leukemia. Early childhood exposure to other children reduces the risk of at least this form of cancer.

We know that immune systems don't develop properly in uninfected or underinfected mammals. And there is evidence that ALL, like some other human leukemias, develops after a viral infection. Children in day care come in contact with a large array of infectious microorganisms during critical periods of immune development. Children who remain at home, especially children in small families, encounter many fewer infectious agents. Because of this difference in the frequency and breadth of infection, it seems likely that the immune systems of these two groups of children develop differently. Such differences could make the underinfected group more susceptible to infection by leukemia-causing viruses, or the immune systems of underinfected children might respond against an infectious virus and then lose control. Since the working cells of the immune system are white blood cells, lack of control during an immune response could lead to dramatic expansion of white blood cells, just like that seen in ALL. Either way, children who encounter fewer unrelated children (infections) during the first few months of life are more likely to develop childhood leukemias. One more strike against sterility.

Allergies, asthma, obesity, inflammation, leukemia—the consequences of good clean living. We humans are complicated creatures.

But we like simple ways of looking at things. After Pasteur showed what anthrax could do to cattle and to humans, bacteria and all the other microscopic vermin became the enemy. After all, along the way, we saw what syphilis, plague, tularemia (named for Tulare County, California, this disease causes terrible ulcers in people's lungs and is one of the most infectious diseases known), and brucellosis (a bacterium that causes abortions in cattle and swine and fatalities in humans) could do. Diseases, we reasoned, especially infectious diseases, are bad. So when Fleming finally brought penicillin to the public in the late 1940s, it seemed a miracle. And as surely as any other miracle, it would have saved my uncle Henry's life, just as it saved thousands of lives, especially among children. We were overjoyed, and rightly so. This was our first and greatest victory in the treatment of infectious diseases. But in our joy, we may have gone too far.

Because of our first successes, and because of our lack of understanding about how bacteria weave the magic that they do, we were all only too eager to start spreading antibiotics everywhere.

At the beginning of the twenty-first century, we have a list of antibacterial drugs as long as our arms. On top of that, we have antifungal, antiviral, and antiparasitic drugs that would drop a billion bugs in an instant. To that we've added antibiotic crib mattresses, countertops, dinnerware, hand soaps, deodorants, mouthwashes, toothpastes, shampoos, toilet cleaners, dish soaps, cleaning sprays, aerosols, powders, scrubbers, toothbrushes, and towelettes. We give antibiotics to our cows and pigs and chickens like there was no tomorrow. Ignoring our four billion years of history with bacteria, we have tried and continue to try to push the bugs out of our lives. And we nearly have.

But not without consequences. Between 1980 and 2005, the number of Americans with asthma rose from 6.7 million to 17.3 million. And the frequency of asthma continues to rise faster than that of any other disease in the United States. By 2020, it is estimated that the number of people with asthma in the United States will nearly double, to 29 million.

Beyond the simple increase in numbers, there are some interesting trends in these increases. The level of parents' education correlates directly with the incidence of allergies. That is, the more educated the parents, the more likely the children are to suffer from allergies. And non-Hispanic white children are more likely than Hispanic children to have had allergies. Which suggests that with money, like education, the more you have the more likely it is that your kids will have allergies.

Though some might consider allergies and asthma as a fair trade-off for infectious diseases, it isn't the use of antibiotics in the treatment of infectious diseases that has led us to our present state. It is the overuse of antibiotics—drugs that we have spread like insulation to deafen us against the threat of infectious diseases—that has brought us to where we now stand.

The more we have done to eliminate the infectious from our lives, the sicker our children have become.

Life and Death and Bacteria

Greg LeMond was once the greatest U.S. cyclist, and before he was done, he was the first non-European to win the Tour de France. Eventually, he won it three times—in 1986, 1989, and 1990. LeMond also won cycling's world championship twice—in 1983 and 1989. Blue-eyed and blond, LeMond was wildly popular with American cyclists, and even the French seemed enamored with him. After three victories in the Tour, American cyclists were beside themselves with joy. But in 1991, LeMond began to falter. That year, even though he wore the leader's yellow jersey briefly during the race, he finished seventh in the tour. In 1992, LeMond failed to finish the race. And finally, in 1994, he announced his retirement from competitive cycling. In the space of two years, LeMond had gone from cycling's greatest athlete to noncompetitive. But it wasn't exactly his fault. A part of him had failed, a part that most athletes give no thought to whatsoever.

The bacteria that cover our skin, our noses, our intestines, and our tracheae are not the only bacteria that we humans need if we are to survive. We also need bacteria inside every one of our cells.

Mammalian cells are eukaryotic cells, meaning they have cellular organelles—little membrane-wrapped structures—inside of them. These include, among others, the Golgi apparatus where proteins are festooned with sugars and directed to their final destinations; the endoplasmic reticulum where proteins are made; the nucleus where the chromosomes work their magic; endosomes where things are assembled, disassembled, eaten, and excreted; and mitochondria. The last of these, the mitochondria, are small peanut-shaped organelles found in varying numbers inside all of our cells. Mitochondria are the source of most of the energy it takes to be a human being. Mitochondria are like little machines that extract enormous amounts of energy from fats and sugars and make that energy into a chemical called adenosine triphosphate (ATP), the stuff of life. Every action that distinguishes a living human being from a dead human being is dependent on ATP. ATP comes from mitochondria. Amazingly, mitochondria are bacteria.

These little bits of us are the descendants of free-living bacteria that long ago moved inside eukaryotic cells. Like all other bacteria, mitochondria have no internal membranes. Mitochondria have their own DNA, but no nucleus, also like all other bacteria. And mitochondria divide when their environment is right, not when their host cell divides. Mitochondria are the living remnants of bacteria tucked away inside every one of our cells. Without these bacteria, we would not be human. Bacteria fuel every human action.

Today, mitochondria's nearest living relatives appear to be *Rickettsia prowazekii*. This is the bacterium that causes typhus, the disease that killed Anne Frank and thousands of others in Nazi concentration camps.

Billions of years ago, the bacteria that would one day become mitochondria were ingested by, infected, or in some other way inserted themselves inside of the primitive ancestors of eukaryotic cells. But instead of destroying one another, these two cells struck a bargain—energy for food. The arrangement was novel by all current standards, but it appealed to both parties. The bacteria found an unlimited supply of food and in return they provided a nearly unlimited amount of energy to their hosts. The face of evolution changed forever.

This sort of relationship is called symbiotic, two different living things inhabiting the same space. And when one of the beings is inside of the other, it is called an endosymbiotic relationship. The one who lives within is called an endosymbiont. Bacteria are endosymbionts that play a big part in our lives.

They also play a big part in our deaths. Besides providing most of the energy we need to be humans, mitochondria regulate how we develop inside our mothers and whether our adult cells live or die. As a human fetus forms, some of its cells must die to allow for normal development. For example, at about week nine of human development, the hands and feet of the fetus look like paddles, because the toes and fingers are webbed. As fetal development continues, the cells that form the webs die. The death of these cells allows for the continued development of normal fingers and toes. The timing for this event is critical. If the cells of the web die too early, fingers and toes don't develop fully. If the cells of the web die too late or not at all, the infant is born with webbed hands or feet or both. The mitochondria inside the cells of the web determine when these cells die.

Similarly, throughout our adult lives, most cells of our bodies must periodically die and be replaced by new cells. The balance between these two events—cell death and cell division—is critical. If cells die too slowly, a person will soon have enough skin or liver for two people, then four, or worse, cancer. If cells die too quickly, organs atrophy and people die. Mitochondria and their products regulate the rate of cell death throughout our lives. And it now appears that just how mitochondria do this also determines just how long and how well we live.

. . .

As a child, Greg LeMond made a list of three things he wished to do. The first was to win the Tour de France. The second was to win a World Cycling Championship. The last was to win the Olympic Men's Pro Road Race.

In 1980, LeMond made the Olympic cycling team but never got the chance to compete, because that year, the Americans boycotted the Moscow Olympics. His professional status kept him from competing later in his career.

In 1983, LeMond won the first of two world championships. In 1986, he won his first Tour de France. His star looked very bright. But in 1987, he was accidentally shot in the back by his brother-in-law while the two men were hunting. LeMond underwent emergency surgery to remove pellets from his liver, his kidneys, his lung, and his intestines. But even after surgery, at least two lead pellets remained in the lining of his heart, and nearly thirty pellets were never removed from the muscles of his back. Nevertheless, in 1989, Greg LeMond overcame all of that and again won the Tour de France. His wounds and the remnants of the shotgun blast could not stand for long between LeMond and the sport he loved most.

But in 1992, for the first time ever, LeMond couldn't finish the Tour de France. It wasn't because of the shotgun pellets. His legs just wouldn't push him up the big hills any longer. And it wasn't because of lack of training or effort. Instead, LeMond was suffering from a disease called mitochondrial myopathy—muscle weakness caused by inadequate energy output from his mitochondria.

We still don't know much about this disease, but one thing was clear even then. When LeMond's bacteria began to fail, he was powerless to change his fate. The mitochondria that had carried him to the pinnacles of cycling were failing him now. The energy that drove his feet through pedal stroke after pedal stroke up Mont Ventoux couldn't do it even one more time. In 1994, LeMond quit the race he loved and retired to Minnesota.

None of us is a single thing, a person apart, an island, a hermit, or a lone wolf. We are orchestras of individuals. And in every race we run, every thought we form, every human victory or defeat, each player has a part.

So too for these ants, strewn across my driveway this morning. The eggs, the queen, the caves filled with drones, and caches of food are rarely seen. But they are there. Without them the rest is meaningless. And though there is a queen, she never gives an order. Nor does anyone else. Yet ants never want for direction. They may be without minds, but they are not mindless.

We are no different. Upon and beneath the surface of our skins, a colony of workers toils continuously. With no apparent connection to our brains, part of us goes about its business tirelessly and mindlessly as ants. We are oceanfuls of life. We are, each of us, as numerous as the stars. No human ever acted alone.

I think at last of Howard Hughes, the great aviator and billionaire who ended his days huddled in a hotel room because of an obsessive fear of infection. No ant would be so foolish. Our salvation doesn't lie between the sterile sheets of a sanitized room. As every ant knows instinctively, it lies instead in the dirt beneath our feet.

3 ●:●

Bugs in Our Genes:
Infection and Human Evolution

This morning, as sunlight falls through her gauze curtains and the whole world smells of purple mustard, Mary Rasmussen, a girl on the verge of becoming a woman, is terrified. She hasn't slept. Perhaps she'll never sleep. Ever since she found it, she hasn't been able for a moment to take her mind off the thing growing between her legs. *This changes everything*, she thinks. And she is right.

Her mind rolls up on itself with fears of cancer and leprosy and homosexuality. But as she touches herself, she doesn't truly believe it is any of those. *No, this is something worse, something much worse. Worse even than cancer.*

Lying in her fleece pajamas, the yellow ones with the dogs on them, Mary brushes her long brown hair from her face and wipes the moisture from her eyes. Not even the ice-white light that falls from her window will hide what Mary knows. And that knowledge spreads like an accusing finger across the screen of her bedroom ceiling, a small fleshy stalk that stands pink where nothing should have stood at all and points toward a dark future. She raises her arm over her head. She smells of soap and fear.

Mary's mother knocks on the bedroom door.

"What is it?" Mary calls out.

"It's time for breakfast and I just wanted to say good morning," Mary's mother answers cheerily through the door, surprised at the tone and timbre of Mary's voice. *She sounds hoarse*, she thinks.

"Okay. You can come in," mutters Mary.

Mary pulls the flowered quilt up to her chin and tries to hide her body beneath the blankets.

Her mother opens the oiled wooden door and walks to Mary's bedside.

"You sound hoarse, dear. Are you okay?" As she speaks, she reaches out and lays her hand on Mary's chest.

Mary knows what her mother is doing. Her mother is looking for breasts. Mary pushes her mother's hand aside.

"I'm fine, Mother," Mary says softly, hoping to sound less hoarse.

Thirteen and still no period, her mother thinks.

"You're sure you're okay?"

"I'm just fine, Mother," Mary lies. But she lies because she has to. She can tell no one, not even her mother, about what is happening. It is too terrible to speak of.

Mary's mother bends over and kisses Mary on the forehead. "Come and get your breakfast, dear."

"All right," Mary responds, though she has no appetite at all. Trying to talk her mother out of breakfast is hopeless.

Mary's mother leaves.

Mary hasn't yet connected her flat chest with the thing between her legs. But she will, and so will her mother.

Months pass, and Mary tells no one of the terror that grows inside of her. But her mother knows Mary still has not started her period. And, as Mary's fourteenth birthday draws near, she insists that Mary see the family doctor.

Mary tries every excuse she can think of to stall her mother. She fakes cramps. She fakes the flu. She fakes homework and exams and dances and deadlines. But her mother doesn't give an inch on this. Finally, the day comes when Mary can think of no more excuses and no way out.

At the clinic, Mary is beside herself. She paces while her mother reads magazines. She peers out the windows. She pretends nausea. Her mother reads magazines. At last the nurse calls for Mary. Her mother begins to follow her into the examining room.

"Mother, I don't need your help with this. I can handle this on my own."

Her mother hesitates. "Dear, I want to make sure everything is okay."

"I'll tell you how everything is, later. Right now, I want you to stay here."

"But . . ."

"Please, Mother. Please stay here and let me do this alone," Mary pleads.

Mary's mother purses her lips and rolls her eyes. Then she gives in and returns to her magazines. Mary looks everywhere for an escape route but finds none. She ends up sitting on the paper-covered examining table wrapped in a thin cotton dressing gown that covers nearly nothing.

She prays. She clutches the gown closer to her and feels the low chill of the building sweep across her pale skin. She chews at her upper lip, annoyed by the hair she finds growing there. She knows her life is nearly over. And it is. But things never ever end quite the way we expect them to.

The doctor and the nurse enter the room.

"Hello, Mary," the doctor says, and the words seem like a sword to Mary. Every inch of her rings with the hard edge of that steel.

The doctor is tall and serious looking and he doesn't wait for a reply. He simply asks Mary to lie back on the table. He feels her neck, under her arms. He opens her gown and listens with his stethoscope to her back and her chest. He closes her gown and lifts her knees. He looks, then feels between Mary's legs. He seems to take forever. Finally, he lowers her gown and checks behind her knees, has her sit up while he strikes her knees with a little rubber mallet.

"Everything seems just fine," the doctor says. "We'll need a little blood to see what is going on inside of you, and that'll be it."

Just like that, and then he crosses the room and leaves. Mary had expected almost anything but this. She expected gasps, cries, maybe indignation or revulsion. But not this.

The nurse comes back into the small examining room with a syringe and a few glass vials.

"We need to draw a little blood, dear."

Mary lifts her arm.

A week later, Mary finds herself in the opposite position—sitting in the waiting room while her mother is closeted in the examining room. Mary hates them both at the moment—her mother and the doctor. It is her body, after all. She has every right to know everything her mother and the doctor know. And she has a right to know it right now. She chews at a piece of nail on her right finger.

In the examining room, the doctor asks Mary's mother to take a seat.

"Your daughter, Mrs. Rasmussen, isn't exactly your daughter," he begins.

"What do you mean? Of course, she's my daughter."

"She is *yours*, certainly," the doctor offers immediately. "That isn't exactly what I meant."

"Then exactly what did you mean?" Mary's mother shoots back at him.

"I meant, she isn't your daughter. She is your son."

Mrs. Rasmussen feels like she has been kicked in the stomach.

"What? Don't be ridiculous," Mrs. Rasmussen gasps.

"He's a boy," the doctor says again, as flatly as if he had said, "I think it might rain."

"That's impossible."

"Actually, it's quite possible. You see, it's a disease, an unusual disease called 5-alpha reductase deficiency. That means that Mary has a mutation in one of her genes. That gene normally produces an enzyme, and that enzyme helps with male development. Testosterone by itself is not enough to make a baby develop into a boy before birth. Another hormone is needed, a hormone called dihydrotestosterone, or DHT. Five-alpha reductase makes DHT from testosterone. Without DHT, boy babies end up looking so much like baby girls that sometimes even doctors and parents get fooled.

"But then at puberty, all that changes. At puberty, when testosterone and other hormones really get rolling, the true nature of the boy comes out.

"Your daughter is developing a penis, and soon she will develop facial hair and other male characteristics if we don't do something."

"Do something?" asks Mrs. Rasmussen, staring at the doctor's belt, unable to lift her head.

"I think, Mrs. Rasmussen, the best course here would be to tell Mary nothing. She thinks she is a girl. In a week or so, we can explain that she needs some minor surgery. Then we can correct her genitalia so that it more closely resembles that of a young woman. After that, we can start a course of hormones that will help with her sexual development in other areas. In short, we can make Mary into the girl you both thought she would always be, and no one but you and I will ever know."

The Power of Genes

The people in this story are made up. The genetic disorder isn't. Babies born with 5-alpha reductase deficiencies (5-ARD) look like baby girls. But when the full force of puberty kicks in, it becomes apparent that these children are boys. Imagine arriving at puberty as a girl and leaving as a boy. I can imagine my nose bigger, my hair blond, my fingers longer, my skin darker. But I cannot imagine suddenly becoming another sex. Sex is a deep, deep part of who I am, maybe the deepest part. Sex is not something I acquired along the way. I—the deep and eternal "I" of me—am male. Always have been, or so it seems.

With all due respect to the "deep and eternal 'I' of me," most of who I am is negotiable. So, personal or not, embarrassing or not, shameful or not, like everything else, even my sex may appear and disappear at the whim of nature. And in this particular case, that whim comes in the form of a gene. A short stretch of DNA that has me in its grip. Whither that gene goes, there goes "I."

Genes lie inside nearly every one of our cells—red blood cells don't have nuclei, so they don't have DNA, but everything else does. What every cell becomes depends, in large part, on the genes it carries. Even though they all carry the same genes, all cells don't end up the same, because each cell type uses different subsets of genes. Just as every piano has the same keys but all songs don't sound alike. Some songs become concerti, some rock, some country, some rap. Some cells become neurons, others lymphocytes, and still others osteoclasts, because a different set of genes rules each cell type. But genes are still behind all of it.

Most genes lie like beads along the strings of our chromosomes—those oddly shaped little ropes that curl inside cells' nuclei. Chromosomes are long strings of DNA plus some proteins and other things to hold it all together. But at the heart of a chromosome is the steady pulse of genes. And those genes direct most all of the processes that make us into the human beings we are.

How do genes do that? How does something so small and so chemical control us?

Genes are our blueprints. Specifically, genes are the blueprints for our proteins. And those proteins make possible everything we humans like to do and even some we may not—like switching sexes at puberty.

Proteins make humans human in two different ways. First, proteins make us look like humans. Bone, cartilage, hair, muscle, nails, corneas, skin, and teeth are protein. How tall we are, what color we are, how our eyes look, how our skin feels, the shapes of our skulls, how our lips taste, and the brightness of our smiles all depend on proteins—proteins that come from genes. And proteins make life happen in one lifetime—drawing a breath, making love, writing a poem, smelling a rose, speaking your name, life. The proteins that make this possible we call enzymes. Enzymes are tiny catalysts, and catalysts make chemistry happen faster. Every chemical reaction that distinguishes a living human from a dead human would occur without any help. It just wouldn't occur in a human lifetime—not a movement, not a glance, not an idea in a lifetime. Enzymes make us happen in real time. Without them we would be nothing more than a bag of very slowly rotting things that might have been. Enzymes come from genes.

Proteins are also antibodies, hormones, transport molecules, and receptors. Being human is all about proteins, and all proteins come from genes.

The Mysteries of the Human Genome

So, if we wish to understand ourselves, we must understand genes.

Because of that, the National Institutes of Health and a private company recently put a huge effort into sequencing the human genome — deciphering the chemical composition of all of the twenty-four human chromosomes. Chromosomes are made of DNA. DNA is made up of four chemicals called bases — adenine, guanine, cytosine, and thymidine, or A, G, C, T. The arrangement of these bases along the chromosomes determines which pieces of DNA are genes and what those genes can do. Sequencing the human genome required reading all of the letters strung out along all of our chromosomes (more than three billion letters total). And the scientists did it. They got them all. Once we stopped patting ourselves on the back and read the words those letters spelled out, we found all sorts of interesting things.

Only about 10 percent of the three billion letters is part of human genes. That adds up to about thirty thousand genes for each of us. Far, far fewer than most of us expected. But even more interesting, most of the DNA inside of human chromosomes isn't human at all. More than half of the DNA in our chromosomes got there as a result of infection. Inside each of us, there is more viral DNA than DNA in human genes. Nearly 8 percent of our DNA is intact viral genomes. Another 40 to 50 percent is viral fragments.

A particular group of viruses, called retroviruses, gave us these genes. Retroviruses are unique among viruses, because retroviruses insert their

DNA into their hosts' chromosomes. We call them retroviruses because they use a special enzyme to do this, an enzyme called reverse transcriptase.

The remnants of these primordial infections we call human endogenous retroviruses or HERVs. "Endogenous" means that we acquired these viruses and viral fragments from our parents, not from a viral infection. Viruses that we acquire through active infections that occur during our own lifetimes we call "exogenous" retroviruses, things like HIV.

Ancient Infections and the Course of Human Evolution

The first of these viral infections appears to have occurred between ten and fifty million years ago—before human beings were even human. Something long-limbed and covered in hair, no thought for the future. Perhaps, this virus crawled in during the trauma of lovemaking or through a wound ripped into one beast by another. Regardless, eventually the virus pushed its way into germ cells—sperm or egg. From there, the virus found the door to the future.

All at once, even from where the first infected ape sat staring from its broad-leafed tree at the vast plain below, the future looked very different.

Over time, this virus worked its way out of the first infected chromosome and spread to others, and from there to still others. Then more viruses followed—another bite, another birth, another ill-fated love—until viral DNA infected many of our chromosomes, and much of our DNA was no longer human, or anything that would become human. Just how much of our ancestral DNA became functional viral DNA is impossible to tell. But as mentioned, today, 8 percent of the DNA found in human chromosomes is functional, potentially active, viral DNA. That is nearly as many viral genes as we have human genes.

The fact that most of the viral DNA that originally infected us is no longer functional doesn't mean that these bits of viral DNA have no effect. Quite the opposite. It is very likely that these viral remnants changed the course of human evolution.

Normally, change among living things occurs very slowly. A cosmic gamma ray strikes a strand of DNA, and a single G becomes a C. As a result of that change, the gene now makes a protein with a glycine molecule where there was once an arginine molecule. The protein's shape and function shift infinitesimally. And, for the moment, that is all. Change as slow as the movements of the continents. Normally.

But the spread of viral infection from chromosome to chromosome changed everything. Suddenly, cataclysmic change was possible. Multiple copies of viral DNA distributed among multiple chromosomes made it possible for chromosomes to rearrange in ways never seen before.

Parts of some chromosomes disappeared, other parts duplicated themselves, and then their genes wandered off down separate paths. Whole pieces of some chromosomes moved into other chromosomes. And with a viral push, human evolution took a giant step into the future.

Investigators at Tufts University have shown that at least 16 percent of the viral elements in the human genome have undergone changes that suggest dramatic alterations of chromosome structure—deletions of large segments of chromosomes, duplications of big pieces of other chromosomes, and DNA shuffling between chromosomes. Such changes undoubtedly altered the course of human evolution.

Once DNA began to rearrange itself in new ways, the rate of evolution accelerated, the paths of evolution multiplied and diverged, the impossible became inevitable. Opportunities for biological change expanded almost as quickly as the heavens themselves. And at the control center sat viruses. Humanity could only watch as the turntable of evolution spun faster and faster—watch and wait, knowing that the wait would be shorter now.

Even the mammalian placenta, the flesh that ties us to our mothers for nine months, was the gift of a retroviral infection.

The development of most mammalian fetuses involves a placenta, a structure that attaches the fetus to the uterine wall during prenatal development. Across the placenta, the mother transfers oxygen, nutrients, and immune factors throughout pregnancy. Most of us take placentas pretty much for granted. But some mammals, like marsupials (kangaroos, wallabies, opossums, koalas, wombats, bandicoots, etc.) and monotremes—egg-laying mammals (duckbilled platypus and the short-beaked echidna)—do not use placentas. Interestingly, in comparison to these other animals, the young of placental mammals develop much faster. This is because of the protective environment of the womb and the transfer of nutrients across the placenta. The result is that placental mammals reproduce more rapidly, a major evolutionary advantage. And the efficient nutrient transfer throughout the development of placental mammals appears to be responsible for a considerable gain in brain development. So much so that even the brightest marsupials cannot compete with some of the dumbest placental mammals.

Placental mammals have a gene called Peg10, which is not found in marsupials. Mice who lack the Peg10 gene do not produce normal placentas, and their fetuses die very early during development. So it appears that Peg10 is essential for the development of a normal mammalian placenta.

Peg10 is derived from a type of gene known as a retrotransposon. Some retrotransposons, like Peg10, are remnants of retroviruses. Peg10 is a modern reminder of an ancient infection. And Peg10 has given to us the wonder of human reproduction and the miracle of the human mind.

From Eve's Rib

Within the men among us, infection has made an especially deep and mysterious mark. The human Y chromosome contains the genes that make men men. Whatever it is that the fetus needs to switch from girl to boy lies within the DNA of the Y chromosome. Three to four million years ago, just as the genus *Homo* was emerging from older primates, a big chunk of the X chromosome popped out and inserted itself into a primitive Y chromosome. That genetic exchange created the human Y chromosome. A virus did that, cleaved off that piece of the Y chromosome. And it was a virus that forced open a stretch of another chromosome so someday men might become men. From the X chromosome's rib, viruses fashioned men.

Original Viral Sin

As I said earlier, most of the bits of viral DNA inside of human chromosomes cannot now function as genes. What were once viral genes are now often missing essential bits of DNA that would allow them to make more DNA, more proteins, and more viruses. Most of our viral genes have died. But not all.

Some still thrive. Inside each of us, a thirty-million-year-old infection still smolders. A virus that infected our ancestors while our ancestors still walked on all fours festers inside each of us. An infection that has raged for millennia in men and women and all those who came before, an infection that will rage for millennia yet to come.

Over time, mutations, translocations, inversions, deletions, and other dramatic chromosomal events alter viral DNA. Usually, these events alter the viral DNA in ways that make it impossible for the viral genes to replicate themselves. The viruses go silent, at least as viruses. These broken bits of viral DNA may alter the course of human evolution, but they will never again make viruses. However, at least one group of viral genes inside of us is still active. By comparing the DNA sequences of several human endogenous retroviruses, investigators in Britain and the Czech Republic showed that at least one family of human endogenous retroviruses still contains infectious genes. This infection has simmered since the Oligocene, since before *Proconsul africanus*—one of our most apelike ancestors—hid in the trees of Africa. For more than thirty million years, longer than humans can imagine, these viruses have been messing with our chromosomes, all of our chromosomes.

Who can say how much we truly owe to viruses?

And viruses are not the only visitors wandering about inside our chromosomes.

The Bacteria in Our Nuclei

When people looked at the sequence of the human genome, they noticed another interesting feature. It appeared that more than 220 of the genes inside of human cells came from bacteria. And these genes didn't seem to have come to us because we had evolved from bacteria. These bacterial genes instead appeared to have been transferred from bacteria into us much later. That is, it appeared that occasionally, during infections, bacteria inserted some of their genes into our chromosomes. Some of those bacterial genes ended up in the chromosomes of our eggs and sperm. So when we reproduced, we passed these bacterial genes onto our sons and daughters.

The scientists determined this by scanning the human DNA sequence for genes that appeared to have come from bacteria. Then to rule out genes we had inherited from bacteria billions of years ago, the investigators looked at the DNA sequences of plants and other animals. The bacterial genes that appeared only in humans, they reasoned, had to have come to us by recent (in evolutionary terms), direct transfer of genes from bacteria into humans. This analysis uncovered 223 genes that appeared to have been transferred to us by bacterial infection.

But there is one other plausible explanation for bacterial genes that appear in humans but not, say, in rats. Since the time rats evolved from ancient bacterial ancestors, the rat genome has undoubtedly lost some DNA—all genomes do. But all genomes don't lose the same bits of DNA or chromosomes. So it is possible rats might have lost bacterial genes that humans still have. That could as easily explain why humans still have some bacterial genes that rats don't. This is called gene loss.

The idea of genes being transferred directly from bacteria to humans is a novel one. No one had predicted anything of the sort. So it is very important to distinguish between gene loss and gene transfer. Gene loss is simply the sort of thing that everyone would have expected based on the known properties of chromosomes and standard Darwinian evolution. Gene transfer is something different altogether, something remarkable.

Since not all genomes have lost the same genes or chromosomes, one way to reduce the misleading effects of gene loss is to compare the genomes of more species of animals. The more animals we compare, the less likely we are to be misled by a lost bacterial gene in one species. If rats don't have a particular bacterial gene but mice and humans do, then it is most likely that the gene came from ancient ancestors of all three species and was lost by rats.

Steven Salzberg and colleagues at the Institute for Genomic Research in Rockville, Maryland, compared the genomes of many more species and found that only about forty of the bacterial genes in the human genome could not be accounted for as ancestral genes. That means the maximum number of genes we may have received recently from bacteria can be no more than forty. And it is possible that comparing the genomes of even more species may further reduce this number.

But even considering all of that, some of the bacterial genes inside the human genome seem to have come to us after we were human beings, or at least after we were *Homo habilis* or *Homo erectus*.

That means that some of the microbes that infected our ancestors left behind the gift that keeps on giving—genes. And as a result, the course that humans followed shifted. One part human, one part virus, one part bacteria.

Living Mosaics

Mitochondria, as we have seen, came to human beings by a long and intricate path that began nearly four billion years ago. Since that first encounter so long ago, mitochondria have given us warmth and motion, speech and hearing, sight and taste and touch, and that tingling feeling we get when we are with the right person. That's a lot we owe to bacteria. But there's more. Mitochondria have been doing something else to living things ever since they came onboard in the Precambrian.

Throughout the eons that mitochondria have been living within us, they have been slipping their genes into our chromosomes. On more than six hundred separate occasions during human evolution, the bacteria that are our mitochondria have inserted pieces of themselves into our chromosomes—the cores of our humanity. Mitochondrial genes now account for 0.016 percent of the human genome. That may not seem like much, but remember, the human genes in our genomes account for only about 10 percent of the total DNA.

For as long as there have been plants and animals, mitochondria have been downsizing their own genomes and moving their excess genes into the nuclei of our cells. Sometimes they have succeeded. Sometimes they have failed. Failures are apparent as pseudogenes—mitochondrial genes that came apart during the transfer. Bits of these genes still reside inside the human genome, but they can no longer function as genes. This, of course, does not mean that these gene pieces have not affected human evolution. Just as with viruses, the pieces of bacterial genes may have disrupted human genes, changed the character of chromosomes, or altered the regulation of any number of human genes. It is impossible today to identify just how mitochondrial pseudogenes may have changed the human genome. But clearly they have.

And pseudogenes are not the only fossilized bits of mitochondria that show up inside animal and plant chromosomes. Real genes, functional pieces of the mitochondrial genome, have crept into other genomes as well. Some of these genes encode mitochondrial proteins. And the presence of these genes inside host nuclei helps the mitochondria to minimize their own genetic efforts as well as ensures a host commitment to mitochondrial survival. What other mitochondrial genes do is less clear. But mitochondrial DNA does not concern itself only with the affairs of mitochondria. Mitochondrial genes and their products also play important roles in human vision, migraine headaches, epilepsy, and cyclic vomiting syndrome (recurrent episodes of rapid-fire vomiting in between varying periods of completely normal health). The roles of these genes in human affairs may be considerable.

But where did the mitochondrial genes inside our nuclei come from? Well, from mitochondria, obviously. But which mitochondria? All of the mitochondria inside of our cells came to us from our mothers—one more way our mothers infect us before they ever touch us. So it might be tempting to guess that the mitochondrial DNA inside our nuclei came from our mothers' mitochondria. But maybe not.

The mitochondrial genes that persist inside human chromosomes must have been transferred from mitochondria to nuclei inside germ cells—that is, eggs or sperm. Otherwise, these genes would never have been passed on to subsequent generations of humans. DNA transfer is more likely to occur with free DNA than it is with intact chromosomal DNA inside living mitochondria. There are a limited number of opportunities for such transfers to happen during the lifetime of human beings. And nearly all of these occur inside the zygote—the newly fertilized egg.

Maternal mitochondria come to the zygote along with the egg. Other mitochondria arrive with the sperm, though many fewer. But soon after fertilization, the egg destroys the paternal mitochondria. So, for a short period of time after fertilization, free paternal mitochondrial DNA floats around inside the zygote. This seems the most likely source of mitochondrial DNA for transfer into human chromosomes. As a result, as the embryo and then the fetus develop, all of the mitochondrial genes inside of mitochondria are maternal, while all of the mitochondrial genes inside of our human chromosomes are probably paternal—an interesting apparent division of labor. But remember, paternal mitochondrial DNA is, in fact, DNA from our paternal grandmothers' mitochondria.

No doubt M. C. Escher would have been intrigued by all of this. Regardless, in spite of the complicated heredity, the inescapable fact remains—bacteria and viruses have irreversibly changed human history. And like the background cosmic radiation that still carries the faint echoes of the big bang, human lives, human behavior, human form, human thought, and human words still reverberate with the peals of microbial bells.

To begin life as a girl and end as a man may seem a miracle or a curse, I suppose, depending on your point of view. But that's the sort of thing that genes do. Miraculous things, horrible things, human things. Canvases full of paint, symphonies full of sounds, the drumbeat of poetry or war. But beneath it all beat the cilia of bacteria and spin the tales of viruses. We are what we are because of those who came before us and those who reside within us. Our pasts and our futures are rotten with infection. The transformation of a girl into a boy becomes less extraordinary than we imagined. All about us are miracles.

4 ⚫⦂⚫

Sepsis and Self-Realization: Knowing and Nurturing Our Infections

The Boy in the Bubble

David Vetter and Carol Ann Dernaret met and married in the mid-sixties. They were very much in love and longed for a family. So, in 1968, they had their first child—a beautiful auburn-haired baby girl. They named her Katherine and she was their pride and joy. The Vetters had few other possessions between them, but they figured somehow they would manage. Unbeknownst to them, they had one more posses-sion than they thought, one they would have gladly parted with if they could. Life went on, and two years later the Vetters had a second child, a boy. They named him David Joseph Vetter III, another handsome baby. But within a few months David began to sicken. The doctors ex-amined his blood and found that he had a serious disease called severe combined immune deficiency (SCID). David's immune system had failed to make T cells, and without T cells David was susceptible to any infection that might come along. By age seven months, those infections killed him.

A short while afterward, the doctors took the Vetters aside and told them that if they were to try again, the odds were only fifty-fifty that another son would have SCID. The gene for SCID is on some X chromosomes. David's mother, because she didn't have SCID, clearly had one X chromosome with the terrible mutation and one without; otherwise, she herself would have died from SCID. Another son would have an equal probability of getting either of Carol Ann's X chromosomes. Therefore, there was a 50 percent chance that the Vetters' next son would be normal. Besides, the doctors told them, if the next son had SCID, the doctors could simply isolate the boy inside a plastic bubble until the technology was found to cure his disease.

Within a few weeks, Carol Ann was again pregnant. Amniocentesis revealed that the Vetters' luck was still running bad. On September 21, 1971, the doctors delivered the second boy by cesarean section and, within seconds, placed the baby, David Phillip Vetter, inside a sterile plastic bubble. David's parents could touch him whenever they wanted—using sealed, sterile, thick rubber gloves, of course.

The physicians had thought of nearly everything. All of David's clothes were sterile, all of his food, all of his toys, and slippers, underwear, toothbrush, toothpaste, soap, towels, and tape. The doctors had thought of everything—everything except David, that is.

By the time David was eight years old he already understood he would never be like others. "Why am I so angry all the time?" he asked his psychologist. "Whatever I do depends on what someone else decides to do. Why school?" he offered. "Why did you make me learn to read? What good will it do? I won't ever be able to do anything anyway. So why? You tell me why?"

David was angry and afraid he was losing his mind. Neither of which, given the circumstances, is difficult to understand. But even without his immune system, even in the face of his fear and his anger, David survived.

Along the way, he grew fascinated with death and fire.

· · ·

Simply acknowledging that human skin and nails and hair and eyes and ears and intestines are crawling with living things isn't enough. Nor is it adequate to understand intellectually that our form, our freedom, and the size of our frontal lobes are the gifts of infections.

To see ourselves clearly, we need another set of eyes.

We have no doubts about the importance of livers and hearts, muscles and immune systems. Humans are not human without these organs and systems. Especially, hearts. Hearts give us love and empathy, regret and remedy.

But we have more DNA from bacteria than we have DNA inside of our hearts, more bacterial cells than heart cells, more bacterial genes than heart genes. Still, we don't imagine that bacteria are nearly as essential to humans as hearts.

That's because we can live, for a while anyway, without our bacteria. But being able to live without something doesn't mean it's dispensable. After all, many people have lived, at least briefly, without their hearts. And David Vetter, though he never chose to, lived without an immune system for twelve years.

The Human Microbiotic System

Open a human physiology text, and here's what you'll find: The human body contains eleven interacting systems: cardiovascular, nervous, immune, musculoskeletal, gastrointestinal, genitourinary, reproductive, integumentary, respiratory, endocrine, and hematologic. All separate from one another and all purely human.

These systems describe a more or less arbitrary division of organs, tissues, and cells into the functional units that make us human. We believe that each of these systems is essential to humanity, because when one of them fails, people change—usually for the worse.

As discussed in Chapter 2, when our microbes fail, we change, also usually for the worse. Mammals without bacteria require 30 percent more water and a third more calories. These same animals are subject to a whole range of infections by microorganisms that don't usually bother infected animals. The intestines of germ-free mice don't develop normally, nor do their immune systems. Uninfected animals often suffer from diarrhea and malnutrition. Animals without bacteria die sooner than infected animals. Uninfected animals develop inflammatory bowel diseases more often than septic animals. Animals have more bacterial DNA than animal DNA, and bacteria control hundreds of animals' genes. Animals treated with antibiotics often develop diarrhea and opportunistic infections. The list goes on. Animals, including humans, without their bacteria are not themselves. Just as animals without immune systems are not themselves. Our infections make us who we are just as much as our immune systems do.

It seems clear, then, that our physiology texts are incomplete. There is a twelfth system throbbing away inside human beings and other mammals, a system we gather during and after birth from our mothers and the world around us, a microbiotic system. A system we cannot see that is as necessary for life as any of the other systems.

For a century or more, our concepts of self have been large enough for only eleven systems. We need to broaden these concepts to include the others who live inside of us—those our parents gave to us, those we took from the dirt, those our lovers left with us, and those we share with one another every day. These, too, are among the things that make us into human beings. These, too, are among the things we cannot live without.

Illness as Imbalance: Wellness as Metaphor

Infection isn't only a way of disease; it is a way of life. Wellness does not mean uninfected. The character and the health of our infections drive our lives. We need those who infect us. To strive for purity would not only be futile, it would be disastrous.

How do you construct a metaphor for that? We haven't had much time. That's the problem with science. It thrusts truths upon us faster than we can construct cages to hold those truths. Machines taught us to think of human bodies as assemblages of cogs and pumps and rotors and wires and switches. I don't know how people thought of themselves before there were machines. But the moment that first piston pushed water through a pipe, men and women changed. The whole universe changed. We became machines, machines at war with a microscopic, biological swarm. For decades, that seemed good enough. And we knew that when we were well, we ran like cotton gins or well-tuned automobile engines. That was health. That was our wish. The swarm was the enemy—the targets of antibiotics and sanitation, vaccination and prayer.

But we aren't machines, not even a little like machines. The swarm is not our enemy, and health has nothing to do with cogs and levers and pistons. We are the swarm. We live in a world rich with germs, a world built by germs. The only way we and others have found to survive in such a world is to welcome germs into our lives, to ourselves become those germs.

We cannot see this world the way it truly is. Our eyes won't allow it. And because we don't see the microscopic frenzy, we forget. But forgetting makes no difference. This world teems with life. This world is life—life so rich and so vital that it covers everything. A pulse so regular that it pushes the Earth itself. Despite our trusted metaphors, we have never known anything clean or pure or pristine or perfect or smooth or sterile or safe. Nor have we ever known solitude.

This world shivers with anticipation, crawls with bugs, and itches with infections. Only those who gave themselves over survived. Only those rank with the smell of infestation made it.

Germs create people, and healthy germs make for healthier people. This may sound simple, but it isn't. Like the man on the tightwire in the slippers, the one who carries the long balance bar and measures out his life in circus steps—it is all about balance.

Infectious wellness requires a healthy, normal flora, a whole series of vital normal infections. Like any other system, the microbiotic system requires nourishment. And like any other system, when malnourished, the microbiotic system struggles and health fails.

Our infections tie us to the world and to one another. Even H. G. Wells, more than a century ago, understood the importance and the security of infection. In his famous novel *War of the Worlds*, it wasn't humans who destroyed the Martians. It was our infections. The same infections that make us forever a part of this world made the invaders forever part of another.

Disease as Disturbance

We may speak of them as though they existed apart from one another, but each of our systems is intimately and intricately intertwined with every other system. A change in kidney function raises or lowers blood pressure. A change in the immune system alters endocrine function. A change in the nervous system interferes with normal gastrointestinal function. And it is no different with human microbiotic systems. Just like the cells in our immune systems, the balance of bugs that live within us and on us has evolved over billions of years. When that balance shifts, so do we.

The interactions between immune systems and the microbes in our intestines are delicate and complex. We have to distinguish the good bug from the bad, destroy the bad, and leave the rest intact. We manage this through some wizardry that involves a continuous low-grade inflammation inside our intestines. This inflammation must be just enough to keep us ready for an unwanted arrival but not so great that it interferes with our essential infections. Just how we manage that remains pretty much of a mystery, but its importance is apparent.

Crohn's disease is a devastating inflammatory bowel disease. A virus, maybe, or a bacterium inadvertently sets off an immune response that the body quickly loses control of. That immune response attacks the bacteria of the gut. This attack causes areas of massive inflammation throughout the large and small intestines, and severe malabsorption, pain, and bloody diarrhea. And then, everything unravels. The colon often ulcerates, and fistulas (small tunnels) may form between loops of intestine or between bowel and bladder or bowel and vagina. The inflamed bowel can no longer absorb proteins or carbohydrates, fats or vitamins as it should. Then there may be arthritis and inflammation of the eyes and mouth, kidney stones, gallstones, and liver disease.

The normal relationship between the gut flora and the host involves a complex network of bacteria-bacteria and bacteria-host interactions, and bacteria-immune system interactions are of special importance. A shift in the balance, a dropped bar, lost footing, and diseases like Crohn's and other inflammatory bowel diseases rise from the once-benign muck to catch our falls.

But the GI system is not the only system affected by imbalances in the gut flora. As I mentioned earlier, there is a clear connection between the development of allergies and asthmas and the mixture of bacteria in the gut. Underinfected people have altered immune systems, and not just in the gut. Though the changes begin there, they quickly spread to other parts of the immune system. And allergies and asthmas are not the only immunological diseases that have their roots in the gut microflora.

Rheumatoid arthritis (RA) is an autoimmune disease driven by an antibody against an antibody. Autoimmune diseases arise when a person's immune system mistakes his or her own body for the enemy and sets out to destroy some or all of that person. Antibodies are proteins made by the immune system. Normally, they float in the plasma looking for intruders. When they find one, they bind to that intruder and deliver it to a white blood cell for destruction. Most of the time, that's a very good thing and helps keep us alive.

But people with RA make antibodies against their own antibodies. Their immune systems have somehow mistaken their proteins for the enemy and aggressively move to destroy that enemy.

This anti-antibody—the one that people with RA make (called rheumatoid factor)—binds to a man or woman's normal antibodies and forms large protein complexes that get stuck in blood-vessel walls and the joints of the knees, wrists, knuckles, and ankles. That causes massive and chronic inflammation. The inflammation destroys joints, cripples, and kills people.

Because this disease begins when a person's immune system attacks his or her own proteins, we call it an autoimmune (*auto* for "self") disease.

What causes a person to make antibodies against his or her own antibodies remains cloudy. Some studies suggest that one important event in the etiology of rheumatoid arthritis is a change in the composition and well-being of the gut bacteria. Just like Crohn's disease, RA is another inflammation burning out of control because of a loss of balance.

And beyond autoimmune diseases, there's a host of problems that can follow antibiotic treatment: allergies and asthmas in children, yeast infections in adults, and so on.

For more than four billion years we've bathed in a broth of bacteria and fungi and viruses and parasites. That's how we made it. But ever since Pasteur first saw the little demons under his microscope, we've been obsessed with eliminating every tiny, crawly, creepy microbug that we could. We have just begun to understand the consequences of that obsession.

Eating Worms: Infection as Therapy

It is November in Tucson. The weather feels more like Colorado in July. El Presidio Plaza hums with life. Dozens of people walk or stand or sit beneath the porcelain sky. Mariachis are playing somewhere near, and the air bubbles with the smells of frijoles and fresh tortillas. A wonderful day. But while the aromas and the sunshine distract us all, something terrible happens, and no one notices.

Each of us revolves in a world limited by the poverty of our own senses. Every day, every moment, a million billion living things pass before our eyes and no one notices.

Near the fountain at the edge of the plaza, a child in a soiled pink-print dress lifts her ice cream cone to her mouth. Just as it reaches her lips, the dollop of frozen cream falls to the pavement onto a spot where another child was just sitting. The little girl reaches with her sticky fingers and lifts the ice cream from the stone to her lips. She brushes aside her auburn hair, and before anyone can do anything about it, she pushes the ice cream into her mouth and swallows. Inside of that clot of cinnamon-pecan swirl, the child's future unfolds its wings and reaches out for her. The child's mother wipes the girl's face and then scolds her for eating from the stone. But the woman is too late. For them both, the world has already shifted its orbit.

A week later the little girl begins vomiting up everything she eats. Diarrhea follows, intense, life-threatening diarrhea.

"She's infected with a rotavirus, a tiny virus shaped like a wheel," says the doctor at the clinic.

"A rotavirus?" asks the worried mother.

"A rotavirus."

"Will antibiotics help?"

"I'm afraid not. Though good hygiene may help to keep the rest of your family safe."

"What can we do for my daughter?"

"I recommend an infection," the doctor says, without missing a beat. "A bacterial infection. A massive bacterial infection."

Each year rotaviruses infect fifty-five thousand children in the United States and more than six hundred thousand children worldwide. In this country, rotavirus infections begin in November in the Southwest and spread across the country and finish in the Northeast around April— much like influenza. Except, without treatment, rotavirus infections frequently kill. Also unlike flu, rotaviruses move from person to person via the oral-fecal route—from feces to mouth to feces and so on. People rarely intentionally eat feces, so the most common routes of infection usually involve events like those described above or unwashed hands and the things such hands have touched.

Rotaviral diseases remain a bit of a mystery. It seems that imbalances in the gut flora allow the virus a space to do its dirty work. Rotavirus-infected children given live *Lactobacilli* have much less severe diarrhea than children given pasteurized (sterile) yogurt. In other words, a change in the composition of the normal floras of these children's intestines made them resistant to rotaviruses. Infection made them whole again.

Just as with any other system in our bodies, among the bacteria in our intestines, balance is the key to well-being. Not purity.

I mentioned earlier that Crohn's disease results, in part, from an imbalance in the gut that leads the immune system to mistake the normal flora for the enemy. In the ensuing attack, the intestine is largely destroyed along with the resident bacteria.

This means that at least one factor in the development of Crohn's and, perhaps, RA is change in the gut flora. If such changes can cause disease, perhaps other changes can cause health. And they do.

Eating *Lactobacilli* appears to improve the prognosis of patients with Crohn's disease. Again, changing the mix of bacteria in the gut changes the course of the disease. And in the case of Crohn's disease, researchers have found an even more astounding treatment.

Crohn's disease, along with other inflammatory bowel diseases, is found almost exclusively among the people of wealthy, developed nations. Nations that, in their obsessions with cleanliness, have left behind another, once constant companion—worms.

Flukes, flatworms, tapeworms, pinworms—slithery, parasitic worms—were for millions of years a part of life for every human being. Almost from the moment of birth to the moment of death (and a few weeks afterward), worms squiggled and squirmed in different parts of our bodies. More than any other single thing, water treatment put a stop to that. The richest source of many worms has always been drinking water contaminated with human waste. That contamination is gone now. Nor do we allow biting insects free reign among our children. These habits have—to our great pleasure—set us, once and for all, apart from worms.

But it turns out we cannot so easily divorce ourselves from a million-year-old marriage. Without our worm partners, our guts just aren't the same.

To make up for this, Joel Weinstock feeds people worms—pig whipworm eggs, to be specific. A while back, Dr. Weinstock, along with a few others, noticed that inflammatory bowel diseases such as Crohn's disease and ulcerative colitis occurred rarely, if at all, among the people of underdeveloped countries. However, among these same people, infections with helminth worms were a way of life. He and others wondered if those two facts might be related. Could it be that we in the developed world suffered from inflammatory bowel disease because we had no worm infections?

Some initial studies in mice suggested that Dr. Weinstock had hit upon something. The next step was people. Same result. Drinking pig whipworm eggs helped all of the first seven Crohn's-disease patients tested. And six of the seven patients went into remission. Drinking pig whipworms cured their inflammatory bowel diseases. Since pig whipworms don't mature fully in people, those intentionally infected are not at risk to develop long-term infections and the complications of these infections. So none of these first patients experienced any significant side effects. Larger trials have since confirmed these original findings. A helminthic miracle. Infection as cure.

And the closer we look, the greater the miracle. Treatment with worm eggs also reduces or eliminates the symptoms of ulcerative colitis, another inflammatory bowel disease with autoimmune roots. And when patients who recovered after treatment stopped swallowing pig eggs, the colitis reappeared.

Other investigators at the University of Iowa found that asthma yielded to a similar recipe of worms. Mice infected with worms showed little or no response to dust mites, egg whites, and other allergens. While worm-free mice reacted fiercely to all of the allergens tested.

Anne Cook and her coworkers at Cambridge University in the United Kingdom showed that extracts of worm eggs prevented type 1 diabetes in susceptible mice. Type 1 diabetes is also an autoimmune disease, in which a person's immune system attacks the beta cells of the pancreas. In normal humans and mice, the beta cells produce insulin. Without treatment, diabetic mice die from lack of insulin and from inflammation—like Crohn's disease and ulcerative colitis.

For most autoimmune diseases, treatments seek only to ameliorate the symptoms, not eliminate the causes—diabetics get insulin, people with rheumatoid arthritis get immune suppressants and anti-inflammatories. But mice whose genetics predispose them to diabetes fail to develop diabetes when injected with worm-egg extracts. Infection, or its equivalent, stops the immune system from making mistakes and that saves mouse lives.

Multiple sclerosis (MS) is another autoimmune disease. In MS, people's immune systems mistake the myelin sheath that wraps around neurons for an infectious enemy. Immune systems then begin relentless attacks against the nerves, especially the nerves of the brain and spinal cord. The results vary a lot from person to person, but MS usually wreaks havoc in the brain and spine. As you can imagine, this damage makes it difficult or impossible for many people with MS to perform even the most routine human tasks.

Murine autoimmune encephalitis is a disease of mice that looks very much like human multiple sclerosis. The nerve damage is similar, and the symptoms look a lot alike. Mice with murine autoimmune encephalitis develop severe hind-limb weaknesses and paralysis.

Unless these mice eat worm eggs.

Eating eggs from either of two different species of worms dramatically reduces both the hind-limb weakness and paralysis in mice with autoimmune encephalitis. Worm wizardry.

Now, because of what mice have shown us, we are beginning to feed worms to humans with MS. It will be months, maybe years, before we know if humans with MS, like mice with autoimmune encephalitis, can be saved by worms. But diseases that have resisted human medicine for as long as there has been medicine are yielding to worms. Worms that crawl through our blood and our lymph, inside of our bodies and out. Worms we have known for longer than any of us can recall. Worms we have curled our lips over and tried our best to rid ourselves of. Worms that may yet save us from ourselves.

Bursting the Bubble

David Vetter did get out of his bubble. The first time was in 1977, when NASA presented David with his own Mobile Biologistical Isolation System—a renovated space suit. David had considerable difficulty getting the suit on and overcoming his fear that the suit might be infected. But once inside, David loved it. He had never walked more than six feet in any direction. Now he walked down the hall for the first time and poured a cup of tea for a nurse. When he handed her the tea, it was the first time David had ever handed anything to anyone.

The last time David left his bubble was in October 1983. His doctors had decided that they should attempt a bone-marrow transplant on David. The T cells in the donor's marrow, they reasoned, would fill the space left in David's immune system by his mutated gene. They performed the procedure immediately.

By December, it was clear that something had gone wrong. David's immune system was not responding as hoped. Fifteen days later, David Vetter died. His immune system could not be repaired, and without it, the world was more than David could handle.

All of David's biological abnormalities and some of his psychological difficulties doctors quickly and easily attributed to his lack of an immune system and his isolation in the plastic bubble. Apparently, no one ever wondered if David's lack of bacteria and other microbes might be at the root of some of his problems.

Humans can survive for a while without a functioning immune system. David Vetter proved that. But no one, not even David Vetter, would argue that a human without an immune system is a normal human being. Humans need their immune systems. Immune systems are part of who we are.

The same is true for our infections.

part two

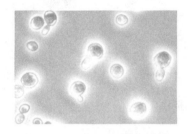

THE LUNATIC
FRINGE

5 ⦂⦂⦂

The Dark Side:
Infectious Diseases

The boy had lost a lot of blood, that much was clear. His skin was ashen and his face hung partly agape from his cheekbone to his chin. A short while ago, a dog had attacked the boy and opened his face like that. He could stand only as long as his mother held him. Now the two of them—the boy and the woman—simply stood and stared at the bearded man in his leather chair. In the smoky air between them hung the question the woman had asked of the French microbiologist.

Pasteur knew that if he did what she asked, he might hang for it. He knew, too, that if he didn't do it, this nine-year-old boy would suffer horribly and probably die. He held his bearded chin in his left hand and pushed his tongue against his front teeth.

The boy, Joseph Meister, stood no taller than the Frenchman's hip. He had been bitten severely and repeatedly. His mother had finally pulled the boy from under the dog, and what they told Pasteur convinced him that it was a rabid dog that bit Joseph. Soon there would be a fever and headaches, then insomnia, hallucinations, excess salivation, twitching, and hydrophobia. After that, if the boy was fortunate, he would die quickly. Pasteur had seen others die from rabies and the sheer horror of that changed the man forever.

Pasteur was not a physician; he wasn't licensed to treat anyone for anything. If he chose to inoculate this boy, it might mean the end of the scientist's career. He looked back at the boy and his mother. There was no time left. Pasteur made up his mind. The brutality of the disease and the fragility of the child pushed him from his chair and into his laboratory. There, a strip of spinal cord hung inside a jar, slowly drying. The cord came from a rabid rabbit killed nearly twelve days ago. Pasteur lifted the strip of meat from the jar and laid it out in a glass dish on the table before him. Then he picked up two scalpels and began to systematically mince the strip of flesh into pieces small enough to inject. Satisfied with the size of the pieces, he added a bit of sterile saline and drew the mixture into a syringe.

Back in his living room, he asked the boy to lie down on the couch. Then, as the boy's mother watched, Pasteur pushed the needle into Joseph's arm, squeezed, and uttered a small but sincere prayer under his breath.

Infectious Disease

At the end of the nineteenth century, hundreds of people died from rabies every year. Almost invariably, these people got the disease from the bite of a domestic animal, usually a dog.

Because of the efforts of Louis Pasteur and many others, it is rare today for someone to die from rabies. And the few cases of rabies that arise each year in this country almost invariably come about as a result of a bite from a wild animal, like a skunk or a bat. The rarity of the disease and the elimination of rabies from domestic animals are due largely to Pasteur's discoveries and intense vaccination programs.

But rabies remains as a glaring example of the pain and suffering that can follow infection. And rabies is only one example of thousands of infectious diseases—diseases caused by prions, viruses, bacteria, fungi, and parasites—that make us sick and kill us. Some infections may be essential. Others are terrifying.

A rhabdovirus virus causes rabies. There are many different types of rhabdoviruses and they cause a whole range of diseases—rabies is one of the worst. The rabies virus is bullet shaped, covered with a large sheet of host-cell membrane, and, fortunately for all of us, is easily destroyed by soaps or drying. Rabies viruses are about 75 nm (nanometers) in diameter and about 150 to 300 nm long. A nanometer is 10^{-9} meters, or one billionth of a meter, far beyond the realm of human vision. These are tiny packets of trouble.

Rabid animals transmit rabies in their saliva. From the bite wound, the virus makes its way into neurons. Then it travels along the neurons to the central nervous system (brain and spinal cord) where it sets up shop. In the central nervous system, the rabies virus begins to destroy neurons as it produces more virus. Some of the new virus particles travel to the salivary gland, where they help spread the disease to others through bites.

In the process of its self-replication, the rabies virus decimates many of the neurons in the central nervous system. The result is a mental meltdown. But the virus doesn't destroy all of the brain—just enough to drive the dog mad, just enough to make the dog want to bite every other living thing it sees and pass on the infection.

Precisely and accurately defining disease is difficult, if not impossible. If we define disease as any significant deviation from normal human biology, then we are all diseased. And perhaps that is so. But such a definition renders the word "disease" more or less useless. If, on the other hand, we seek some narrower definition of disease, it becomes arbitrary. How much someone must differ from the physiological norm before we call him or her diseased is clearly in the eye of the beholder.

Nevertheless, none of us would have any difficulty deciding to call someone with symptomatic rabies diseased. The hallucinations, howling, extreme pain, and madness that often follow a rabies virus infection eliminate all need for splitting hairs.

Besides the complicated task of definition, disease clearly involves many factors. Even with a disease that seems as obvious as rabies, not everyone infected with the same virus has the same symptoms. In fact, some of the infected have no symptoms whatsoever.

Infection alone is rarely, if ever, sufficient to produce disease. Many other elements determine the fate of the infected. And though we might think that the infected are the only ones affected by the path and the outcome of infectious diseases, we would be wrong.

But it isn't exactly our fault. Our visions of the invisible, you see, are blurred because of the darkness of the caves where we first found microscopic life.

Fear and Loathing in the Microbiotic Underworld

One of Pasteur's greatest claims to fame is that he was the first to realize that microscopic living things caused many things, including diseases. He arrived at that conclusion in a most remarkable way.

Hold your left hand up to a mirror. In the mirror you see another hand. The hand in the mirror has the same number of fingers and thumbs as the hand at the end of your arm. But the flesh-and-blood hand and the mirror hand cannot be rotated so that the hands face the same direction (e.g. palm up) and all the fingers and the thumbs of the two hands line up exactly. The hand in the mirror looks like a right hand and the one at the end of your left arm looks like a left hand. That's called chirality. Chiral things have nonsuperimposable mirror images—like the left hand at the end of your wrist and the "right" hand in the mirror.

Amino acid and sugar molecules are also chiral. That is, both amino acids and sugars exist in either of two forms, right-handed or left-handed. A chemist can easily validate this by observing what happens to a beam of light when it passes through crystals of amino acids or sugars. Depending on the character of the crystals, the light beam will rotate either slightly to the right (right-handed) or slightly to the left (left-handed) or not at all when there are equal mixtures of both right- and left-handed molecules present.

Laboratory-made amino acids don't rotate light at all. Amino acids synthesized by chemists contain equal mixtures of both right- and left-handed molecules. But amino acids and sugars made by plants and animals rotate light to the left or to the right, never both. Something about life favors one over the other.

Naturally occurring amino acids are always left-handed, even those found inside meteorites that have wandered billions of miles before crashing into the Earth. Naturally occurring sugars are always right-handed. Nobody has ever figured out why that is, why both sugars and amino acids naturally occur in only one chiral form.

Regardless, this simple fact unlocked a door for Pasteur, a door locked by the mysteries of chemistry. Behind that door lay a clue to the origins of life, the nature of disease, and the magic of fermentation.

Tartaric acid is an organic acid made by plants, including grapes. Like amino acids, tartaric acid comes in both left- and right-handed forms. Pasteur noticed that the tartaric acid found in the lab had no effect on a beam of light. But tartaric acid from fermented grapes (wine) rotated the angle of a light beam to the left. To test his theory that living things caused the fermentation of wine, Pasteur used mold to ferment a mixture of left- and right-handed tartaric acid. The fermentation produced only left-handed tartaric acid. To Pasteur, this proved that living creatures must have been responsible for the fermentation that produced the wine from the grapes, not the process of dying, as others had proposed.

From that, he made the astonishing intellectual leap to the belief that microscopic life caused disease. Once again, he was right.

Several wonderful things followed: controlled fermentation of wine, beer, cheese, and bread dough. A cure for France's ailing silk industry—the silkworms were suffering from infections that limited their output of the raw material. Vaccines for anthrax and for rabies. Three of Pasteur's five children had died of typhoid fever, a disease caused by a bacterium. He was highly motivated to understand and treat infectious diseases. In the end, his insights and his contributions were among the most remarkable of all modern science.

However, there was a downside. Microbes might be fine for fermentation, but inside of human beings, microscopic life caused diseases — horrible diseases, like rabies. Pasteur had proved that. And from that time forward, we have always thought of microorganisms as the enemy and health as a war against that enemy.

Most bacteria don't sicken or kill people.

It appears that on and in and above and below this Earth there are about 10^{29} bacteria. Clearly, most of them do us no harm, and a significant few even do us great good. But even if 99.99999999999999 percent of bacteria are harmless or benefit and save their hosts, that still leaves 10^{15} bacteria to waste their time killing someone. That's one with fifteen zeros after it. That's still a million bacteria for every single person living here. A million bacteria that can quickly become several billion bacteria once they find the right host. It takes only a few — by bacterial standards a very, very few — rogue bacteria to change the whole picture, to change human lives, to destroy livers and fingers, hearts and minds, bones, blood, marrow, and muscle.

Most microbes are no threat to humans, which is good — but not good enough.

Infectious Diseases: The Not-So-Final Frontier

The World Health Organization (WHO) is responsible for monitoring the health of Earth's peoples. Part of this job requires this organization to evaluate the trends of the present to foresee a little of the future. Here is what they have to say about infectious diseases:

> An infectious disease crisis of global proportions is today threatening hard-won gains in health and life expectancy. Infectious diseases are now the world's biggest killer of children and young adults. They ac-

count for more than 13 million deaths a year—one in two deaths in developing countries.

Over the next hour alone, 1,500 people will die from an infectious disease—over half of them children under five. Of the rest, most will be working-age adults—many of them breadwinners and parents. Both are vital age groups that countries can ill afford to lose.

Infectious diseases remain the leading cause of death in the developing countries and a leading cause of illness in the developed nations. Tens of millions of people die each year from infections, but only six different types of infectious diseases account for more than 90 percent of all those deaths. According to the World Health Organization, in 2002, 3.9 million people died from lower-respiratory tract infections, 2.8 million from HIV/AIDS, 1.8 million from diarrheal diseases, 1.6 million from tuberculosis, 1.2 million from malaria, and 0.6 million from measles.

Lower-Respiratory Tract Infections

The upper-respiratory tract of humans is always infected. Every breath we draw holds a fistful of bacteria and viruses and fungi. Those creatures settle out on the cells that line our noses and throats and our sinuses. Occasionally, this results in diseases, like colds. But usually we never even notice that our noses or mouths are home to other living things. Lower-respiratory tracts are different. These parts of our bodies usually house few if any strangers. Human trachea (windpipe), bronchi (where the trachea forks and leads into the lungs), and lungs themselves are pretty much sterile. Occupation of these tissues by viruses or bacteria, even for relatively brief periods of time, can lead to serious problems.

Viruses are the most common cause of lower-respiratory-tract infections. This is probably because viruses are also the most common cause of upper-respiratory-tract infections—colds, flu, etc. In severe or chronic upper-respiratory-tract infections, some of the viruses manage to find their way beyond the lungs' primary lines of defense and into the lower-respiratory tract.

Viruses that find their way into the lower reaches of human lungs meet fewer challenges and often quickly infect the cells of the lungs. Once inside, the viruses begin to replicate themselves, and as they do, human cells begin to die. Dead cells cause local inflammation.

Then the blood vessels expand. Fluid moves out of the blood and into the infected tissues. The infected person begins to cough. His chest hurts. Breaths come with effort that sounds like a steam locomotive. Without treatment, the infection spreads. And as each infected cell buds off more and more virus, or worse, as cells explode, the virus moves quickly from cell to cell. The disease worsens.

Now the damage the virus has done opens the door for bacteria that have been hanging around forever in the upper-respiratory tract waiting for just this moment. *Streptococcus pneumoniae* from throats, *Staphylococcus aureus* from noses—a normal part of our microbiotic systems—move quickly into the lungs. A new garden begins to sprout in spots that should be barren. Alveoli, the little sacs where toxic carbon dioxide is exchanged for life-giving oxygen, become clotted with dead white blood cells. The infected, with a sound that defies metaphor, begin to cough up green or brown phlegm.

The inflammation worsens, the walls of the lungs swell, and pneumonia rises from the mucus. Large chunks of lung throb with inflammation. Whole fields of alveoli shut down. Breathing becomes labored. Still the infection spreads. More lung is lost. And finally, if the patient isn't treated, he suffocates in his own mucus.

Or maybe there is no virus, and it all begins when one child coughs in another's face. The spittle, rich with *Bordetella pertussis*—the agent of whooping cough—leaps between the two children. If the second child is unvaccinated, the process begins again—a new bacterium, the same old pneumonia. Or it's parainfluenza virus, or influenza virus, or adenovirus. Then there's inflammation, swollen blood vessels, a pool of mucus, pneumonia, and death.

As bad as this sounds, these horror stories describe only a few of the causes of lower-respiratory-tract infections we understand. But lower-respiratory-tract infections are the third leading cause of death world-wide and the leading cause of death from infectious diseases. Nearly four million people die of lower-respiratory-tract infections each year. We are only able to identify the infectious agent in a small number of these individuals. Some of the others surely die from known causes, but most of them don't. We have no idea what is killing most of these millions every year. Whatever it is, it's contagious. That much we know.

HIV/AIDS

An instant, a single intimate instant or a bloody needle, and a door opens onto a new way of living and a new way of dying. A virus leaps from one human to another and the world changes.

Chapter 14 goes into a great deal more detail about HIV and AIDS. Here, I consider only the outlines of this global disaster.

Human immunodeficiency virus (HIV) probably arose inside a chimpanzee, maybe as early as the 1950s. For thousands of years, humans have fed on chimps. Somewhere along the line, one of the hunted bit one of the hunters. When their blood mixed, the virus leapt between the species—and a new story began, one without a clear beginning, one without a happy ending. And that story will reach us all within a decade or two, possibly three.

The human immunodeficiency virus differs from most of the viruses that infect people. HIV is a retrovirus; like those other viruses that have infected us for millennia and left their footprints in the sands of human DNA—the HERVs—it inserts copies of its genes into our DNA.

Usually, HIV infects humans during one of our most sacred acts—lovemaking. A thing we were born for, a thing we have been preparing for for more than three billion years. A thing we, as a race, cannot live without. HIV has coupled sex and death. It is difficult to imagine a ploy more sinister or more likely to succeed.

Once a person is exposed to HIV, the virus infects white blood cells—lymphocytes, most notably—and a few others. Inside of lymphocytes, HIV strips off its coat and inserts its DNA into human chromosomes. Then it does nothing, or nearly nothing, for ten or fifteen years or more. But one day, immunity wanes, and then the virus rises like the phoenix from our genetic ashes and claims all it finds. As it does, the last of our immune systems crumble. And suddenly, we stand naked, defenseless, broken.

The world moves in. Oral thrush—a yeast infection of our mouths—begins to fester. Kaposi's sarcomas—virus-caused tumors almost never seen in healthy humans—rise like clumps of mushrooms. *Pneumocystis carinii*, a fungus, begins to spread its spidery tendrils inside of our lungs. Herpes infections that have been dormant for years erupt across our mouths and our anuses. Lymphomas, more virus-caused tumors, blossom in our brains. To be stripped of immunity is finally to understand the onslaught that each of us escapes every day, and to see our deaths in every mirror.

But before that vision rises, we have time, years even, to love whom we love and to touch those whose touch we cannot live without. Time, too, to spread the virus that will one day kill us and those we have loved.

Nearly three million deaths per year, and rising.

Diarrhea

Diarrhea? Diarrhea isn't something you die from, is it? Annoying, certainly. Inconvenient, embarrassing, foul-smelling, painful, putrid, I'll grant you. But lethal? Hardly.

Lethal? Certainly, especially in children.

Diarrhea is the leading cause of childhood death in developing countries—again the horror of being infected and poor.

Each day, completely unnoticed by most of us, people secrete and reclaim more than four gallons of water. Our stomachs and our small intestines secrete most of that water. Large intestines reclaim it. Anything that interferes with this process can make life difficult or, especially if you're poor, impossible.

A healthy man of 150 pounds has a total blood volume of about five quarts. A person with cholera can lose up to thirteen quarts of water a day. Then the blood quickly begins to lose water to other tissues. Brain functions change. Blood vessels collapse. Normal metabolism grinds to a halt. People die.

Campylobacter jejuni, a bacterium, is the most common cause of diarrheas in the world and the second most common cause of food-related diarrheas in the Western world. *Salmonella* is the second most common cause worldwide and the most common cause of food poisoning in the United States. And in flood- or war-ravaged countries, *Vibrio cholerae*, another bacterium, is often a major contributor to diarrheal diseases and death.

C. jejuni infections begin when a person drinks contaminated water or eats contaminated food. The ingested bacteria attach to the cells in the jejunum (the middle section of the small intestine), the ileum (the terminal section of the small intestine), and the colon. Then, as they divide and grow, the bacteria stimulate an inflammatory response. Remember, the colon, or large intestine, normally absorbs about thirteen quarts of water a day and returns it to the circulatory and gastrointestinal systems. Inflamed colons don't do this very well. As a result, the water remains in the intestine and causes severe diarrhea.

For a short period of time, at least, infected people shed large numbers of bacteria in their diarrhea. Often the runny stools find their way into the groundwater and then into the drinking water. If water treatment is inadequate, or the person's hygiene is poor, the disease quickly moves to others.

The diseases caused by *Salmonella* or *Vibrio cholerae* are similar to that of *Campylobacter*. All are transmitted by ingestion of water or food contaminated with feces from or bodies of infected people. All can cause severe diarrhea. And, under the right circumstances, all can kill.

And do, nearly two million times a year—mostly young, mostly poor.

Tuberculosis

In 1820, when the poet John Keats coughed a spot of bright-red blood on his handkerchief, he told a friend, "It is arterial blood. I cannot be deceived. That drop of blood is my death warrant. I must die." Less than a year later, Keats was dead. He was only twenty-five years old. Before the century was out, millions of others followed, including Frédéric Chopin, Anton Chekhov, Robert Louis Stevenson, and Emily Brontë.

Just before the end of the nineteenth century, tuberculosis was responsible for one in seven deaths worldwide. And if you considered only the premature deaths of the middle-aged, tuberculosis took one in three. Clearly, a devastating disease.

But the beauty of some of those it killed made tuberculosis seem "a disease of the young, pure, and passionate." In Alexandre Dumas's 1852 novel, Camille dies coughing up consumptive blood. Consumption also claims Mimi in Giacomo Puccini's 1896 opera, La Bohème. Scenes where it seemed love itself had ruined the lovers' lungs.

But it wasn't thwarted love at all. It was a bacterium, a mycobacterium, that turned Camille's lungs to cheese and curds, robbed Mimi of her breath, and suffocated them all.

In 1892, Robert Koch first showed that tuberculosis was caused by *Mycobacterium tuberculosis*. Koch infected guinea pigs with tissue taken from brains and lungs of tubercular apes and humans who had died from bloodborne tuberculosis. In every case, the guinea pigs developed identical diseases. And in each pig, Koch found the same organisms—mycobacteria. From these experiments came the proof that *M. tuberculosis* caused tuberculosis. And from this work also came Koch's postulates for establishing the cause of an infectious disease:

1. The specific organism should be shown to be present in all cases of animals suffering from a specific disease but should not be found in healthy animals.

2. The specific microorganism should be isolated from the diseased animal and grown in pure culture on artificial laboratory media.

3. This freshly isolated microorganism, when inoculated into a healthy laboratory animal, should cause the same disease seen in the original animal.

4. The microorganism should be reisolated in pure culture from the experimental infection.

Even today, these postulates remain the gold standard for proof of the cause of an infectious disease.

Tuberculosis begins when a healthy person inhales *M. tuberculosis*–laden droplets coughed up by another person with the disease. Once inside the lungs, the mycobacteria travel to the alveoli, where they are eaten by large white blood cells called macrophages. With most bacteria, the macrophages kill the cells they have eaten. But not with tuberculosis. *M. tuberculosis* has developed the ability to block the macrophages' killing mechanisms. So even inside the macrophages, the bacteria continue to divide.

A few bacteria migrate out of the lungs and travel via the bloodstream to other parts of the body. In some places where the bacteria settle—like the brain, bones, lungs, and kidneys—they may cause further disease. In any of these tissues, bacterial replication and inflammation can quickly lead to severe symptoms and sometimes death. The course of the disease depends a lot on the immune response of the infected man or woman. Some people can eliminate the mycobacteria before they cause serious disease. Others cannot. Either way, it is ultimately the growth and division of the mycobacteria that cause tuberculosis.

So, in the late 1940s, when antibiotics became available, tuberculosis began to decline. Antibiotics killed the mycobacterium that caused tuberculosis and halted the whole disease process. Another triumph for antibiotics. And in 1954, when a fairly effective vaccine was found, deaths from tuberculosis tumbled even farther. Tuberculosis became a disease of the 1800s, mostly. And though there was nothing romantic about tuberculosis, it was no longer an issue among healthy men and women of the developed world. That left the rest of us free to remember it any way we wished to. In her 1978 book *Illness as Metaphor*, Susan Sontag wrote of tuberculosis as a thing mostly in the past. She spoke accurately and movingly about the romantic perception of the disease in the late 1800s and early 1900s. But like all the rest of us, she never imagined that tuberculosis wasn't done with us. Not nearly.

From 1944 on, every year the number of deaths from tuberculosis continued to decline. Until about 1986.

In 1986, tuberculosis turned on us, and there was a sudden increase in the numbers of people dying from this disease, the first such increase in more than three decades. Since 1986, the number of deaths from tuberculosis has continued to rise. In 1986, something changed, and it wasn't us.

Suddenly, we were dealing with a different bacterium. The same antibiotics that had been so successful in treating tuberculosis now did nothing. Vaccines that had once seemed to help became useless. There were no rules anymore. In 2000, 1.6 million people died of tuberculosis. In 2004, more than two million died.

Between about 1950 (when the antibiotic streptomycin became available) and 1986, *M. tuberculosis* had learned a trick or two. Now, when the bug ran up against our best antibiotics, nothing happened. The bacterium had figured out a way around our biggest guns, and it began again to regularly kill men and women. Genetic mutations— little changes accumulating in the DNA every time *M. tuberculosis* divided—made that possible. Once enough of those small changes had accumulated, the mycobacterium was no longer quite the same. Then, when it encountered streptomycin, *M. tuberculosis* simply shrugged and went on about its business. This process of change isn't unique to tuberculosis. It happens even in humans, only much more slowly. Chapter 6 explores this more thoroughly.

For the moment, it is enough to realize that through nothing more than the natural and inevitable process of mutation, tuberculosis has risen from its ruins and once again threatens us.

Malaria

It came out of central Africa and opened a road all the way to England. Along the way, it probably contributed to the fall of the Roman Empire and killed Alexander the Great. Certainly it killed Oliver Cromwell, created the gin and tonic, and changed the course of human history. By the middle of the twenty-first century, it will sicken and kill many more of us.

Chapter 11 describes the modern crisis of malaria in much greater detail. This section provides an overview.

"Bad air," Dr. Francesco Torti called it. Or *malaria*, in the physician's native Italian. Torti named the disease malaria because he believed evil swamp gases were responsible for the ailment. The *dottore* was wrong, but the name stuck.

Protozoan parasites *Plasmodium falciparum, P. vivax, P. ovale*, or *P. malariae* cause malaria. *P. falciparum* causes the worst form of malaria.

Working among Roman ruins that date to A.D. 450, American archaeologists found malaria genes in the bones of buried children. And likely the disease was already old by the time it found its way into those children. Probably, malaria began in central Africa and worked its way north from there. The parasites now infect somewhere between three hundred and five hundred million people worldwide. And today in Africa, a child dies from malaria every thirty seconds. There have been outbreaks of malaria as far north as Great Britain and the northern United States.

Mosquitoes, in particular *Anopheles* mosquitoes, transmit malaria. The cycle begins when a female *Anopheles* mosquito bites an infected human. First the mosquito pierces the skin and secretes a small amount of saliva into the wound to prevent the blood from clotting. Then she feeds. As the mosquito feeds, she draws blood—along with malaria parasite—into her mouth, and then into her gut. In the gut of the mosquito, the parasites mature and some of them infect the mosquito's salivary gland. When the mosquito feeds again, she first injects saliva. In that process, she passes the parasite along to a new host.

Once inside again, the malaria parasite migrates to the liver and begins to grow and divide. As the infected cells release parasites, the malaria enters the bloodstream and infects red blood cells. Inside the red blood cells, malaria parasites divide rapidly, so rapidly, in fact, that the red blood cells burst. When that happens, more parasites flood the blood and infect more red blood cells. The rupture of the red blood cells also releases toxins into the blood, and those toxins cause the fever and chills, aches and pains that are characteristic of malaria. Quickly, the blood fills with parasite, so that when a mosquito bites the infected person, parasites will certainly find their way into the mosquito and on to a new host.

Treatments for malaria are various and variably effective. The British were among the first to attempt to prevent infection and to try to slow the disease among the infected. When the British colonized India in the late 1800s, they were confronted with the problem of endemic malaria. It was already known that quinine could help prevent infection, but it had proved nearly impossible to get British troops to drink the bitter medicine. To make the drink more palatable, the officers added gin to quinine-rich tonic water, and for good measure they threw in a dash of lime juice to add a little vitamin C and help slow the scurvy. The result was, of course, the gin and tonic, still one of the world's most popular drinks. Many remedies for malaria still rely on quinine and related chemicals, but none of these remedies has proved entirely successful. So efforts have focused instead on eliminating malaria.

Attempts to eradicate the malaria parasite or the disease-bearing mosquitoes have been successful in some developed countries, but not at all in Africa—where the incidence of the disease is highest. Administrative problems, mosquitoes resistant to insecticides, parasites resistant to antimalarial drugs, and poverty have all stood between Africa and eradication of malaria.

Measles

Wait. People don't get measles anymore, do they?

In 2000, the Centers for Disease Control and Prevention reported roughly 30 million cases of measles worldwide, and about 777,000 deaths that year. Worldwide that year, measles was the fifth leading cause of death in children. In 2003, in the United States, about 150 people got measles and two of them died.

People do get measles. But mostly, it's poor people.

A morbillivirus causes measles. This family of viruses is responsible for several diseases, including measles in humans and distemper in dogs. Measles is one of the most contagious diseases known. It spreads between humans when someone with measles coughs or sneezes and releases thousands of droplets of saliva and mucus thick with measles virus. Once inhaled, the virus infects the epithelial cells of the upper-respiratory tract. From there, it spreads to the eyes, the gut, the urinary tract, the lymphatic system, the blood vessels, the brain, and the spinal cord. A week to two weeks after infection, the boy or girl begins to cough, scratch his or her eyes, and find bright lights painful. Another seventy-two hours and the unmistakable rash appears as the child begins to run a fever. The rash is caused by the immune system's attack on infected epithelial cells surrounding blood vessels.

If everything goes right, in a few days the rash disappears, the fever subsides, and the child is immune for life.

If everything doesn't go right, the virus spreads through the lungs and causes viral pneumonia—an inflammation that causes 60 percent of the deaths from measles virus infection (fever gets most of the rest of the children). Or even worse, the virus spreads through the brain and causes a viral encephalitis—inflammation of the brain—a very dangerous disease.

Measles is preventable. An effective measles vaccine has been available for almost forty years. But it costs money—not a lot of money, but for the truly impoverished, it is beyond reach.

More than half a million children died from measles in 2003, the sixth leading cause of deaths due to infectious diseases worldwide.

Louis Pasteur changed the look of this world forever. His germ theory, his microscopes, his vaccines, his pasteurizations, and his work with silkworms showed us that what we see with our eyes is the tiniest fraction of what is truly in front of us all the time.

At age forty-five, Pasteur suffered a severe stroke. His work in the laboratory ceased. He couldn't see well enough. He couldn't control his limbs any longer. He couldn't think as clearly as he once had. In his journal he wrote:

> There is a time in every man's life when he looks to his God, when he looks at his life, when he wonders how he will be remembered. It can happen with age or with tragedy or closeness of death. I am lying here at age 45, not able to feel my left side. Not knowing if this stroke that has befallen me will end my life before the mysteries that I have unlocked can be resolved. I have asked God throughout my life to be able to ". . . bring a little stone to the frail and ill assured edifice of our knowledge of those deep mysteries of Life and Death where all our intellects have so lamentably failed."

Pasteur never recovered the use of his left hand or leg, but that didn't slow his work. It was after his stroke that Pasteur completed his work on fermentation, cholera, anthrax, and silkworms. It was after the body-splitting stroke that Pasteur began his work on rabies. And it was years after his stroke that Louis Pasteur found the resolve to vaccinate Joseph Meister.

Pasteur's injections saved Joseph Meister's life. After being injected thirteen times with the minced rabbit spinal cord, the boy made a complete recovery. Hundreds of others followed.

Pasteur died in 1895. The French government awarded him hero status and buried him in the Cathedral of Notre-Dame. Later, those who most understood Pasteur moved his remains to the Pasteur Institute, the research establishment built for him in honor of his work on rabies.

Joseph Meister returned to the Pasteur Institute before Pasteur's death. For years, Meister served as the gatekeeper there. But after Pasteur died, Joseph also became the guardian of the great scientist's crypt, a job he asked for because of his debt to Pasteur.

In 1940, the Nazis occupied the Pasteur Institute. As part of that occupation, the Germans wished to open Pasteur's tomb, and they forced Joseph Meister to help. Because of his love of Pasteur and his shame, forty-five years after Pasteur saved Meister's life, Joseph ended that life with his own hands.

Human diseases come in many forms.

6 ⁘

Taking a Turn for the Worse: The Deteriorating State of the World

Snow falls. Pieces of sky come to earth, smothering the trees, even the leafless trees, the garbage at the curb, the brown grass, the dog droppings on the lawn, a single dead jay. As I watch, the snow soaks up every stain. Only two degrees beyond this thin sheet of glass—two degrees of Fahrenheit temperature to warm this morning world and slow the creep of ice.

Today the luxury of shelter has the rough feel of necessity. The air itself looks frozen as each sweet crystal falls six-armed and innocent. Discarded by the spent storm, they lie in heaps beyond the blue pulse of my computer screen, behind my house, in my drive, on my neighbor's steeply pitched roof.

Everything smells of absolutely nothing, the cold too fierce for smell, too solid for sound. One after another, the ice steals my senses, until all that remains is the pure white glare emptied of context.

The storm spits and swirls and comes to ground. The ground disintegrates. There is only snow and the bony fingers of winter trees. No risk. The snow has buried risk.

It would taste of hot days at sea. It would feel like glass.
Cold enough to kill.
I am so easily fooled.

In the 1970s, we thought we had beaten infectious diseases. Polio, measles, mumps, pertussis, tetanus—one after another had fallen before the scythe of vaccination. Bacterial infections, like strep throat and staph, were no match for the powerful antibiotics we had found. A dozen more of humankind's archenemies crumbled under the hammer of modern medicine.

So certain were we that in 1967—led by our surgeon general William H. Stewart—we shifted medical funding away from infectious diseases and focused our attention on chronic illnesses: cancer, heart disease, stroke.

Like the snow beyond my window, the powdery fall of antibiotics gradually covered the horror of infectious diseases, lulled us, as it gathered, into thinking we were safe forever. We gave no thought to what we might find when the spring came and stripped the ice from the frozen ground.

Fewer than twenty years later, we finally looked back beneath the snow of our hubris. From where we stood, it was clear that forever had lasted less than two decades.

We might have seen it sooner if we had ever chosen to look beyond our own borders. But even global predictions were rosy. In 1978, the United Nations member states wrote and signed the "Health for All 2000" accord. That agreement "predicted that even the poorest nations would undergo a health transition before the millennium, whereby infectious diseases would no longer pose a major danger to human health. [And] as recently as 1996, a World Bank/World Health Organization–sponsored study by Christopher J. L. Murray and Alan D. Lopez projected a dramatic reduction in the infectious diseases threat [worldwide]."

Deaths from infectious diseases, particularly tuberculosis and pneumonia, declined more than 8 percent each year from 1938 to 1952. By the mid-1960s, pertussis (whooping cough), polio, smallpox, tetanus, and diphtheria vaccines were widely available. As a result, illnesses and deaths from numerous infectious diseases dropped drastically. For example, in the United States, the number of cases of paralytic poliomyelitis dropped from more than fifty-seven thousand in 1952 to only seventy-two in 1965, and the last case of smallpox in the United States occurred in 1949.

The scales had grown thick across our eyes.

Between 1973 and 1998, while we looked the other way, twenty familiar diseases, including cholera, malaria, and tuberculosis, reemerged or spread dramatically. And at least thirty new diseases crawled out from under the blanket we had thrown to cover infectious diseases. We still have not found a cure for even one of these new diseases.

Pathogenic Microbes and the Diseases They Cause, Identified Since 1973			
YEAR	MICROBE	TYPE	DISEASE
1973	Rotavirus	Virus	Infantile diarrhea
1977	Ebola virus	Virus	Acute hemorrhagic fever
1977	*Legionella pneumophila*	Bacterium	Legionnaires' disease
1980	Human T-lymphotropic virus I (HTLV 1)	Virus	T-cell lymphoma/leukemia
1981	Toxin-producing *Staphylococcus aureus*	Bacterium	Toxic shock syndrome

YEAR	MICROBE	TYPE	DISEASE
1982	*Escherichia coli* 0157:H7	Bacterium	Hemorrhagic colitis; hemolytic uremic syndrome
1982	*Borrelia burgdorferi*	Bacterium	Lyme disease
1983	Human Immunodeficiency Virus (HIV)	Virus	Acquired Immuno-Deficiency Syndrome (AIDS)
1983	*Helicobacter pylori*	Bacterium	Peptic ulcer disease
1989	Hepatitis C	Virus	Parentally transmitted non-A, non-B liver infection
1992	*Vibrio cholerae* 0139	Bacterium	New strain associated with epidemic cholera
1993	Hantavirus	Virus	Adult respiratory distress syndrome
1994	Cryptosporidium	Protozoa	Enteric disease
1995	Ehrlichiosis	Bacterium	Severe arthritis?
1996	nvCJD	Prion	New variant Creutzfeldt-Jakob disease
1997	H5N1	Virus	Influenza
1999	Nipah	Virus	Severe encephalitis

Source: U.S. Institute of Medicine, 1997; WHO, 1999.

And this was before severe-acute respiratory syndrome (SARS) snaked out of Asia and the Marburg and dengue viruses reemerged. Despite predictions to the contrary, infectious diseases remain the leading cause of death in the developing nations.

While our heads were turned, viruses, bacteria, and parasites — representatives from every major class of infectious agent — re-created themselves and arose, seemingly from nowhere, to strike at us once more.

Add to that the appearance of antibiotic-resistant strains of staphylococci, tuberculosis, malaria, and others, and it is clear that what we left for dead beneath a blizzard of antibiotics and vaccines in the 1970s was very much alive and well in the 1990s. Even in the United States, the number of deaths from infectious diseases has risen from a historic low of about 87,000 in 1980 to more than 170,000 in 1998. The number of people carrying hepatitis C virus — a cause of liver cancer and cirrhosis — has increased to four million. And even though, because of multidrug therapies, fewer people in the developed world are dying from AIDS, the number of those infected worldwide continues to spiral.

How did that happen? What did we miss?

Nearly everything.

Our Static View of a Changing World

Most of what happens around us goes completely unnoticed. It is too small, or too fast, or invisible to our eyes, or too soft, or too low, or too high, or too far away, or it has been going on for as long as there have been human beings — so no one notices, or it is too near, or it just takes too long for a human to notice during one measly lifetime.

We humans have short attention spans, not to mention relatively short life spans. Because of that, we miss a lot of what happens around us, especially those things that happen very slowly. Some butterflies live for only a few days. Imagine one of those butterflies perched on the limb of an aspen tree that might live for more than a million years. Because nearly nothing about the aspen tree changed during the butter-fly's lifetime, the butterfly might easily assume the aspen tree was dead.

Like the butterfly on the aspen, humans often just don't have enough time to see what is actually happening, to discriminate between the living and the dead, to notice the might and the persistence of the microscopic.

A billion microbes, a gamma ray, a radio wave, ultraviolet or infrared light, X-rays, an elephant's subsonic song, the death of a star, the breath of a butterfly, the crunch of ants' feet as they march, the sound of starlight, the rise or fall of mountains, the lazy sound of meandering rivers we miss because of our blindness or deafness, or age, or size, or lack of interest. We notice only the tiniest fraction of what happens in our world.

Human beings have been around for two million years or so, but we have known about microbes for only about 0.01 percent of that time. Perhaps that's part of it. But another part of our mistake with microbes was arrogance. How could anything so small stand in our way? Surely, bacteria, smaller than ideas themselves, were no match for human beings, were certainly less of a threat than others who had dared endanger our lives—wolves or grizzly bears, snow leopards or panthers.

A single invention—gunpowder—eliminated, or nearly eliminated, all of those animals. So it might make sense, at some level, to imagine that a single invention—like antibiotics—would be sufficient to eliminate bacteria. Or a single process—like vaccination—might be adequate to disarm things as small as viruses or bacteria. And that together, vaccination and antibiotics could rid the world of its greatest curse. A world without bacteria was no more difficult to imagine than a world without saber-toothed cats.

It was, after all, the 1970s. We had walked on the moon, seen quasars, split atoms, put them back together, dropped them on Japan, seen San Francisco collapse, made a baby in a test tube, created radios and televisions, mapped the seafloors and then drilled holes in them, and dammed the Colorado River.

Perhaps it was just too hard for the people who had changed this world so much to take a thing as small as a bacterium seriously. Regardless, we all turned our backs on bacteria and looked for other challenges. Thinking all the while of change, but never believing in it.

Human Change

According to "The Global Infectious Disease Threat and Its Implications for the United States," prepared in 2000 for the CIA: "With few exceptions, the resurgence of the infectious disease threat is due as much to dramatic changes in human behavior and broader social, economic, and technological developments as to mutations in pathogens."

I am not certain I fully agree with that statement, but it does make a point. Even if we missed the nearly invisible changes taking place in the microbial world, you'd think we might have noticed what macroscopic human beings were up to. And during the past three decades, we've been up to a lot.

Whole cities and nations have grown up and collapsed. Airline travel, rail travel, and automobiles have continued to speed the movement of people from place to place. Overpopulation has crippled health-care initiatives and driven people from farms into overcrowded cities and poverty. Technology has filled our lives with materials and methods no one could have predicted. Our forests and our wetlands have mostly shrunk, and our cities have reached out like tentacled beasts into what was once farmland. Only change remained constant.

Nothing was static. Not us, or any of the rest of the world. Sometimes, though, change is hard to see when you're standing right in the middle of it.

Human Movement and Human Disease

The movement of whole societies from agrarian to urban lifestyles brings people into more frequent contact with each other—an obvious opportunity for the agents of infectious disease. Often, dramatic changes in income and the quality of food, sanitation, and housing follow such migrations.

Several studies have proved that the persistence of many infectious diseases depends on a certain population density. With too few people, schools, churches, football games, and movies just don't bring many people together. Fewer sneezes get traded, fewer coughs are swapped, and fewer kisses are given or received. Diseases do arise in small groups of people, and sometimes whole communities die. But often it simply ends there. A few or a few hundred people dead. But the next group of people lives too far off to be scorched by the fire.

The opposite is true with large mobile societies. We live in crowded cities where we brush up against one another nearly every moment. Each of us regularly comes in contact with hundreds, if not thousands, of other people. We move about the planet and between isolated populations of people with an ease once reserved for birds.

Abruptly, epidemics become pandemics, local outbreaks spread like wildfires across mountains and oceans and continents. Our own conveniences have cleared away many of the barriers that stood between us and infection.

There is also the sudden forced dislocation of large numbers of people, such as occurred in the former Yugoslavia, or in the Sudan, India, Sri Lanka, Cambodia, Laos, Rwanda, and the American Gulf Coast. Suddenly, thousands, maybe hundreds of thousands of people must live together in areas ill equipped to handle one-tenth that number. Sanitation breaks down and the water becomes filthy with human waste and mosquitoes and flies. Diseases that thrive on poor sanitation—like cholera and malaria—or diseases that thrive on human contact—like HIV/AIDS and tuberculosis—spread quickly.

As of 1998, more than 120 million people lived outside the country of their birth. Millions more relocate every year. Along with them come the scourges of humankind.

Human travel and commerce move millions more potentially infected people from place to place at near the speed of sound. Every person shipped by air from Europe to the United States, or from the United States to Japan, or from anyplace to anyplace is a potential carrier. It was air travel that spread SARS from China to Singapore to Canada in a matter of days. A few years before, a multidrug-resistant strain of *Streptococcus pneumoniae* (which is capable of causing a severe and sometimes fatal pneumonia as well as meningitis) spread from Spain to the rest of the world in a matter of weeks.

Besides any infected paying passengers, planes carry stowaway rats and mice and mosquitoes and flies and fleas and ticks around the world every day. All of these creatures carry diseases that can also infect humans—plague, onchocerciasis (a blinding parasitic infection also known as river blindness), elephantiasis (a worm that plugs up lymphatics), malaria, Lyme disease, hantavirus, and God only knows how many more.

Sometimes by choice, but much more often by force, we are a mobile species. Those who feed upon us feed upon that mobility.

Human Behavior and Human Disease

It is difficult to say whether human behavior has changed dramatically over the past century or that a greater variety of human behavior has become more visible simply because there are more human beings. I suspect the latter. Regardless, certain human behaviors spread diseases.

One very old, very frequent, and very necessary human behavior is now spreading disease, as it has for millennia. Sexual intercourse is hardly something new. Few of us would be here without it (remember the test-tube babies). But recently sex has gained a lot of additional attention.

HIV/AIDS moves between people most often during the act of sex. Anal intercourse transmits HIV infections most effectively, because this form of intercourse is much more likely than vaginal intercourse to result in bleeding. HIV hides inside of lymphocytes, a type of white blood cell. The best way to transmit HIV is to inject whole virus or infected lymphocytes directly into the blood of an uninfected person.

Human semen is loaded with whole virus and lymphocytes. In an HIV-infected man, semen contains millions upon millions of viruses. Sexual intercourse often involves the transfer of semen, and when sexual intercourse results in bleeding, lymphocytes in semen gain rapid access to the uninfected partner's blood. More than half of the time that results in a new infection.

But heterosexual intercourse is also effective at transmitting HIV. More than 80 percent of the AIDS patients in the world today live in sub-Saharan Africa. Most of their infections resulted from heterosexual contact.

Besides HIV, syphilis, gonorrhea, genital herpes, genital warts, hepatitis B and C, and probably several other infectious agents move from person to person during sexual intercourse. Sex—the simple act of human coupling that is essential to our futures—is soaked with risks.

Lastly, intravenous drug use contributes significantly to the spread of many diseases such as HIV/AIDS and hepatitis B and C. Injection of any human pathogen directly into the bloodstream transmits the disease very effectively. Blood is full of essential nutrients, proteins, carbohydrates, nucleic acids, and lipids. There is no broth any richer than the one that pulses inside of men and women. Microorganisms love it. Skin forms a major barrier to infection. But open it up and inoculate what's underneath, and bacteria and viruses and fungi and parasites grow like weeds.

Intravenous administration of drugs is a relatively new phenomenon, at least as new as needles and syringes. Human drug addiction is likely much older. But with the introduction of needles, drug addiction became a means not only for relieving human pain but also for sharing human disease, a very efficient means.

Human Achievement and Human Disease

As already mentioned, ships, planes, trains, trucks, and automobiles carry lots of infected passengers—human and nonhuman—throughout the world. Planes, trains, trucks, and automobiles also move food throughout the world—millions of tons of it. That shift in international commerce has changed the way foodborne illnesses spread. An outbreak of salmonella in a slaughterhouse for chickens in China is no longer just China's problem. An occurrence of bovine spongiform encephalopathy (mad cow disease) in Canada is no longer just Canada's problem. The globalization of food suppliers and the rapid shipment of food across national boundaries can quickly spread foodborne disease.

Dam building has increased mosquito habitat. Hospitals have become prime sites for acquiring staph infections. Surgery and many other medical procedures place patients at high risk of infection. While technological advances have improved our lives in some ways, these same advances have opened new avenues for disease and infection.

As we change how we medicate, the places we build, even the ways we think, the character of infectious diseases will change, too. Microscopic life dominates this planet like no other, and that is because microscopic life is more adaptable than any other.

Microbial Change

Everything discussed so far has to do with human acts and migrations—short-term shifts that alter the way we interact with bacteria or viruses or insects or worms. But humans are not the only things that are changing. Even if we stayed in the same places, ate the same foods, and breathed the same air, the course of infectious diseases would meander. That's because the bugs themselves are agents of change.

If you examine the numbers of people sickened or killed by influenza virus over the past, say, twenty years, what you see is a pattern that looks like the teeth of a saw. Every two or three years, there's a tooth—a spike in the numbers of people who were ill with or died from the flu. In between each of the teeth, you'll find a valley where most people, even if infected by influenza, noticed little, felt fine, and lived happy and productive lives.

Why is that? What's happening out there that creates those saw teeth?

The Tao of Influenza

When you or I inhale someone else's influenza-laden sneeze, we open ourselves up for disease. The influenza falls first on epithelial cells—the outermost cells that line our noses and our tracheae. Normally, a layer of mucus protects these cells from infection. The influenza virus has evolved a special molecule (neuraminidase) to strip that mucous layer. Once the mucus is out of the way, influenza uses another molecule (hemagglutinin) to bind to our epithelial cells and inject itself inside those cells. Once inside, the virus strips off its outer coat and releases its RNA. The RNA takes over the cells' reproductive machinery and makes thousands of copies of itself, then makes thousands more viruses. These new viruses bud out through our cells' membranes and infect more cells. Some of these infected cells die and inflammation follows—noses run, eyes itch, and we sneeze and spread the disease to others.

In a normal healthy person, some of the flu virus is captured by specialized cells and presented to the immune system. The immune system immediately swings into action to try to limit the damage from the infection. But that takes a while, maybe a week or two. So for a week or two, we feel lousy, drip like a faucet, and sneeze a lot. But finally, for most of us, our immune responses reach high enough levels that we begin to destroy the virus-infected cells. The infection ebbs, the symptoms decrease, and we begin to resemble something human again.

Then our immune systems do something nearly miraculous. They dedicate a few million cells to the memory of the influenza virus. We set those cells aside for future skirmishes. The next time we inhale a friend's sneeze, because of the cells we stowed away, the immune response reaches protective levels much faster. It's like having all the firefighters in place before the fire starts. As the flames begin to lick at the roof of our mouths, our antibodies and immune cells suffocate the fire so quickly the flu never burns hot enough to make us sick.

That explains the troughs between the saw teeth. The flu comes along, makes many of us sick, kills a few of us. But those who survive are immune to further attacks. The flu may try, but quick as it does, our immune systems rise up to smack it down.

Over the millennia, influenza has evolved a way around this. Faced with the awesome power of mammalian and avian immune systems, influenza began to adapt in two ways.

First, influenza discovered ways to accumulate mutations, changes in its RNA that caused changes in the shape and substance of the virus. Over time, these changes became sufficient to fool human immune systems, even humans immune to influenza. Like the criminal who, after months of plastic surgery, walks right past security guards who've seen him a hundred times before, influenza ignores our defenses and marches into our lungs. A change of face that fools everyone. Then the number of the sick and dying rises. Again human immune systems kick in, and after a while, the flu is gone. For a year or two, we bask in the warmth of wellness, and then the whole process begins again.

As if accumulating mutations weren't fast enough, influenza viruses have found an even more dramatic way to change their looks. Sometimes, at the same moment, two types of influenza viruses infect a single cell. Inside such cells, after the influenza viruses strip off their outer coats and expose their genes, they can exchange whole segments of RNA. Then the genetic alterations are abrupt and sweeping. A virus that once infected only chickens can now infect humans. A virus that once infected humans now has a completely new face and a whole new wardrobe with which to cloak itself. Nothing is as it was, and all the immune responses that had saved humans or pigs or chickens or ducks are impotent. Everyone is suddenly at risk again. Once more the teeth of the saw begin to rise. Once more people and animals begin to die.

And it is not just influenza viruses that have evolved the means to escape immune responses. Some parasites can regularly and repeatedly change their appearances enough to fool an immune system. Other viruses interfere with activation of immune systems. Some bacteria also interfere with our immune responses. Still other viruses, like HIV and equine infectious anemia virus, have built-in mechanisms for the accumulation of mutations. As these viruses replicate their nucleic acids, the viruses systematically introduce errors that cause alterations in their proteins. Some, maybe most, of these changes destroy the virus. But a few do not. A few make subtle changes in the shapes of the proteins, and the virus remains just as infectious and lethal, sometimes more so. Then these changed proteins slip past old immunity. Even during a single infection with HIV, the virus that is present late in the infection is not the same virus that initiated the infection. Immune systems usually cannot keep up.

When threatened by an immune response, herpes viruses escape into our neurons, and they can hide there for a human lifetime. Then, one day when we get too much sun or too much stress or simply grow old, the viruses reappear—crawl back down the long arms of our neurons—and reinfect our lips or our genitalia and blossom once more into ulcers where we can spread the viruses from person to person. Nothing we've found so far can prevent this process.

Mycobacteria lodge inside our white blood cells, the most powerful weapons in our first lines of defense against bacteria. That's like an enemy soldier walking in to our military command center. And inside those cells, the mycobacteria sabotage our destructive machinery and set up house.

Adenoviruses, Epstein-Barr virus, and human cytomegalovirus all have learned how to avoid activating the immune system. Myxoma viruses, cowpox, and other herpes viruses have found ways to prevent their host cells from being killed by the missiles of the immune response. And vaccinia virus and hepatitis B virus can circumvent the action of some interferons—powerful agents that normally slow or stop the progression of many virus infections.

The Untimely Demise of Antibiotics

Microbes have figured out more than just how to fool immune systems. Antibiotics, once the sharpest swords in the battle against infectious diseases—especially bacterial diseases—have fallen, one by one, against the onslaught of bacterial evolution.

More than 80 percent of us have had *Staphylococci aureus* in our noses or on our bodies or in our vaginas at some point in our lives. About 30 percent of us carry *Staphylococci aureus* around with us continuously. Most people never even notice that they have been colonized by *S. aureus*, unless it gets into their blood. When that happens, the bacterium begins to grow unchecked in the nutrient-rich blood. As it grows, *S. aureus* can produce any number of toxins, some of which can kill. Toxic shock syndrome is a graphic example of the devastating effects of these toxins.

Toxic shock syndrome was virtually unknown before about 1979. Though tampons had been around since 1936 (invented by a physician in Denver, Colorado), in 1979, Procter & Gamble introduced the Rely tampon. This tampon was the first made using polyacrylate fibers, and it was superabsorbent. Procter & Gamble's motto was, "Rely, it even absorbs the worry." At that time, there were still no government regulations regarding the content or manufacture of tampons. The superabsorbency of the Rely tampons created two problems. First, women believed they could leave them in for longer periods of time. This often dried out the vaginal walls and allowed the tampons to swell to considerable size. During removal, the swollen tampons could easily damage the vaginal walls. And because of the polyacrylate fibers and their superabsorbency, the Rely tampons eliminated the exchange of air and fluid between the inside and outside, creating a low-oxygen environment inside women's vaginas. *Staphylococcus aureus*—a normal resident of the human vagina—is a facultative anaerobe. That means that this bacterium thrives with low or no oxygen.

Low oxygen allowed for overgrowth of the *Staphylococci*. The damage to the vaginal walls caused by the tampons opened the door for these bacteria to move from the vaginal vault into the bloodstream.

By the beginning of 1980, physicians had diagnosed more than twelve hundred cases of toxic shock syndrome. In these women's vaginas there were dramatic overgrowths of S. *aureus*. As it grows, S. *aureus* produces hemolysins, leukocidins, an enterotoxin, exfoliative toxins, and toxic-shock-syndrome toxin. The exfoliative toxins disrupt the epidermal layer and cause massive blistering, often followed by extensive scarring. Toxic-shock-syndrome toxin causes systemic activation of lymphocytes and macrophages; that activation causes systemic inflammation. Massive vasodilation (swelling of the blood vessels) leads to a dramatic drop in blood pressure and vascular collapse followed by shock and multiple-organ failure. Early in 1980, Procter & Gamble pulled the Rely tampon from store shelves, and fewer than six hundred cases of toxic shock appeared in 1981. In 1985, manufacturers removed polyacrylate from tampons, and the incidence of toxic shock fell below three hundred.

Nevertheless, because of our close association with this bacterium, S. *aureus* remains the leading cause of gram-positive sepsis (bacterial infection of the blood) in the United States. Most often these infections involve hospital-derived strains of staph.

Penicillin, discovered in the 1920s by Alexander Fleming, binds to and blocks an enzyme that some bacteria (including S. *aureus*) need to assemble their cell walls. Without cell walls, bacteria are helpless bags of protoplasm that quickly succumb to osmotic imbalances and carnivorous host cells. When Fleming first found penicillin, the antibiotic destroyed nearly every strain of S. *aureus*. But by the 1950s, only a decade after doctors began to use penicillin widely, most S. *aureus* strains in the United States were resistant. American drug companies responded by developing other antibiotics, especially methicillin, to treat staph infections. Around the same time, researchers also discovered the antibiotic vancomycin. But because of some significant side effects, this antibiotic wasn't as popular as methicillin. Within a few years, methicillin-resistant strains of S. *aureus* began to appear. This was especially true in areas like Detroit, where intravenous drug users regularly included methicillin along with heroin in the hopes of preventing infections from dirty needles. From the streets, these new strains of staph moved into homes and hospitals, or vice versa. By 1990, the bacteria isolated from 15 percent of patients contained staph that was resistant to methicillin. And 22 percent of the staph samples isolated from critical-care patients were unfazed by methicillin. Things were changing. By 2000, more than 55 percent of staph isolated from critical-care patients was resistant to methicillin. Unsettling, certainly. But we still had vancomycin.

Vancomycin *always* worked against staph. Some doctors believed (or at least desperately hoped) that S. *aureus* would never find a way around vancomycin.

But in Japan in 1997, a strain of S. *aureus* appeared that required much more vancomycin than normal to stop the infection. Later that same year, another partially vancomycin-resistant strain of staph appeared in Dearborn, Michigan. S. *aureus* wasn't waiting for anyone. The future was writing itself across the walls of our hospitals, but no one seemed to notice.

In 2002, the future exploded. More than a year before that, a forty-year-old woman had come to a clinic in Detroit because of pain, swelling, and redness in her foot. The doctors had thrown antibiotic after antibiotic at the infection, and nothing had worked. In April 2002, the doctors finally gave up and started amputating toes. After the surgery, a thorough examination of the amputated tissue uncovered exactly what everyone had pretended could never be: vancomycin-resistant *Staphylococcus aureus*. Our last line of defense had been breached. And it appeared that staph had accomplished this by gathering up the pieces of other bacteria. In a relatively short time, this new threat had picked up everything it needed to beat us at our own game.

Just two months later, another case of vancomycin-resistant staph appeared in Michigan.

In June 2002, Dr. Wayne Brown, a Wayne State University professor, received a swab taken from a woman with an infected sore. Early analysis suggested that the sample contained S. *aureus*, but the cultures were showing strong resistance to vancomycin. Dr. Brown thought that unlikely.

But when he repeated the tests, he got astounding levels of vancomycin resistance. Immediately, he called the Centers for Disease Control and Prevention in Atlanta.

Within two weeks, the CDC had a team on site in Michigan and was screening other patients for the antibiotic-resistant staph.

Like last year's thistle seeds buried beneath the snow, the bacteria were beginning to rise, and spring was not very far off.

But as Ron Iverson, an emergency room physician in Casper, Wyoming, told me, "Try telling a frantic mother with a screaming two-year-old at three in the morning that it is best not to treat the boy's ear infection with antibiotics." You don't usually get very far.

Some of us understand that there is a problem. It's just that when push comes to shove, we don't think it is *our* problem.

Mutations, swapping genes—nothing stands still.

S. *aureus* isn't unique. It doesn't mutate faster, swap genes faster, or evolve faster than other bacteria. What we have witnessed with S. *aureus* is happening with many other microbes.

Evolution has given us antibiotic-resistant strains of tuberculosis, parasites, and fungi, as well as antiviral-resistant viruses. We swing at them, and they swing back. We are aiming at moving targets, and sometimes they move surprisingly fast. In the laboratory, it is possible to develop antibiotic-resistant strains of bacteria within days. Because bacteria divide so rapidly—sometimes as often as once every twenty minutes—bacteria evolve very quickly. At each cell division, there is the opportunity for mutations to be introduced and/or replicated. And each of those mutations has the potential to confer resistance to one more antibiotic. On top of that, bacteria swap genetic information. So it is possible for a normal, old strain of bacteria to become antibiotic-resistant literally overnight.

The Present

The combination of all of the factors considered here as well as those in the table that follows has led to some dramatic changes in the character of infectious diseases.

Factors Contributing to Infectious Disease Reemergence and Associated Diseases	
CONTRIBUTING FACTOR(S)	ASSOCIATED INFECTIOUS DISEASES
Human demographics and behavior	Dengue/dengue hemorrhagic fever, sexually transmitted diseases, giardiasis
Technology and industry	Toxic shock syndrome, nosocomial (hospital-acquired) infections, hemorrhagic colitis/hemolytic uremic syndrome
Economic development and land use	Lyme disease, malaria, plague, rabies, yellow fever, Rift Valley fever, schistosomiasis
International travel and commerce	Malaria, cholera, pneumococcal pneumonia
Microbial adaptation and change	Influenza, HIV/AIDS, malaria, *Staphylococcus aureus* infections
Breakdown of public health measures	Rabies, tuberculosis, trench fever, diphtheria, whooping cough (pertussis), cholera
Climate change	Malaria, dengue, cholera, yellow fever
Source: Adapted from U.S. Institute of Medicine, 1997.	

Humans are intimately tied to infectious microorganisms. It is impossible for a human move to go unnoticed or uninfected in this bacterial and viral world.

This isn't a war, it's a waltz—we lead, they follow, then they lead, and we follow.

For decades we have focused on eliminating the threat and sterilizing the world. That isn't possible. The world we rely on is infectious. Perhaps the only cure for our illness lies inside another metaphor.

Outside, the snow has stopped falling. The air has cleared. It is cold still, like the fist of the sea where this water came from. The sky is back, hard blue rubble. Beneath, snow covers everything—all is clarity and purity. And it will remain so for a while, perhaps a long while. But beneath the snow, beneath the leaf rot and broken branches, under the animal droppings and the frozen corpses, the earth crawls with microscopic life, and the living will have their way.

7 ⣞

The Occult:
The Hidden Face of Human
Disease

Jenny Moore was a child prodigy. She had dark eyes that spoke of thoughts far beyond her twelve years. And even at four feet eleven inches tall, she commanded respect. She had a flair for the dramatic as well. She loved to have her audience seated in the piano room, while she waited in her bedroom. Once all were in place, her father would tap lightly on her door and leave. Then Jenny, often in taffeta, would walk solemnly to the piano, pull out the bench, and sit. While straightening her dress, she would look at her audience, but only for moment. Too much contact with the people could break her fusion with the music. No one had to ask for quiet. Jenny pushed her hair from her high forehead.

And then came the magic—Rachmaninoff's Piano Concerto no. 3 in D Minor, Liszt's Hungarian Rhapsody no. 2, and Chopin. People cried.

This day, as on most days, her concert went perfectly—except today she missed the final note, just ever so slightly missed it. No one but Jenny noticed. But she knew immediately that something was very wrong. And she suspected, even then, that that missed note was the beginning of something different.

Jenny acknowledged the applause, but she was shaken. Her father could see that.

During the two years that followed, Jenny could do nothing but watch as multiple sclerosis (MS) stripped each of her gifts from her. First it was her fingers and the tremors that rose as the disease peeled the insulation from the nerves in her brain and spinal cord. The notes no longer came where or when they should. Rachmaninoff was lost, then the ballades, then even "Für Elise" left her.

The muscles in her legs faltered. Her father had to carry her to the piano. At the keys, she would stare deeply into the gloss of the black wooden panel and cry. Even that ended when her disease blinded her.

Her neck began to tilt at an odd and often painful angle. Her mouth fell into a permanent grimace. She wet herself.

At age twelve, Jenny could play all of Rachmaninoff and all of Chopin's ballades. By age fourteen, she was dead.

Jenny isn't her real name. There are still people who might be uncomfortable hearing her story retold. Just the same, Jenny died of MS. Back then, no one suspected that it might have been an infection that killed her. But thoughts about MS—like so many other diseases that seemingly have nothing to do with infection—have changed.

Multiple sclerosis is a remarkable disease. Few are destroyed as quickly as Jenny was, but all suffer. Beneath the tremors, the incontinence, and the pain burns the white fire of immunity. Immune systems are weapons of mass destruction. They are the only reason we are here. Immune systems battle infections that would otherwise eat us alive. That's what they have evolved to do, and they are very good at it, usually.

Occasionally, immune systems make mistakes. These are often only tiny mistakes, but the consequences can be devastating. In multiple sclerosis, an immune system mistakes the myelin sheath (the insulation that surrounds certain nerves in the brain and spinal cord) for the enemy. Systematically and relentlessly, the immune system then begins to pull a person's nerves apart. The immune self attacks the neurological self—with a vengeance.

The Dark Side of the Moon: Multiple Sclerosis

What would make an immune system do something like that? We don't yet know. But as with other serial killers, we have some interesting clues. MS strikes women about five times more often than men. And among women, those of Northern European descent are most susceptible. Lots of arguments have been offered to explain this, but none has achieved true respectability. About one in seven thousand people in the United States has MS. And then there is the oddest fact about MS: People living above the fortieth parallel north and below the fortieth parallel south are more than twice as likely to develop MS as people living between these two parallels.

The fortieth parallel north, in this country, runs from about Newport News, Virginia, along the northern borders of North Carolina and near Fort Collins, Colorado, and Provo, Utah, and almost right through Chico, California. People north of this line are significantly more likely to develop MS—except for native peoples. Native peoples north of the fortieth parallel in North America and south of the fortieth parallel in Australia and New Zealand rarely get multiple sclerosis. And some groups, such as the Inuit of North America, appear to be completely immune to MS.

Interestingly and inexplicably, women who grow up in Colorado are five times more likely to develop MS than women living in other northern states and ten times more likely than women living in southern states. But, as with all other northern states, children under the age of fifteen who move to Colorado become as likely as those born in Colorado to contract MS. People who move to Colorado after age fifteen are no more likely than their southern neighbors to contract MS.

Occasionally, there have been outbreaks of MS, and groups of people have simultaneously developed the disease. Scientists call such outbreaks clusters. Between 1943 and 1989, four clusters of MS occurred in the Faroe Islands. This group of islands rises between Iceland and Scandinavia and was occupied by British troops during World War II. After that occupation, the number of cases of MS increased each year for twenty years.

Public health officials have reported other MS clusters among people living together in one neighborhood or people working in the same office, in the United States and Europe. And a friend of mine who has MS told me that a surprising number of people from the neighborhood where she grew up have been diagnosed with MS. One MS cluster occurred in 1986 in Galion, Ohio—population 12,391. In that year, twenty-five cases of MS were diagnosed, about twice what might be expected. Nothing about Galion's past or present suggested an obvious source for this "outbreak" of MS. Except one: In 1960, an old graveyard was dug up to make way for a new school. Since a gymnasium would occupy the space where the soil had been, the dirt was offered free to anyone who would haul it away. Many of the citizens of Galion picked up a load or two of that earth and hauled it to their homes.

We call this disease multiple sclerosis, because when we examine people with MS—using an MRI, a PET scan, or at autopsy—we find multiple white areas that look like scarring in their brain and spinal cord—thus, multiple sclerosis (*sclerosis* comes originally from the Greek word for "hardening"). These areas in the brain and spinal cord are white because something has stripped off the nerves' myelin sheath. Myelin is a fatty substance that, in a sense, insulates the nerves and causes faster conduction of nerve signals along certain nerves.

Not only is the disappearance of myelin responsible for the name of the disease, it is also responsible for the symptoms of MS. We call diseases that result from loss of myelin sheath demyelinating diseases. There are many such diseases, but MS is one of the most common in humans.

The fact that women, and especially women of Northern European descent, are more susceptible to MS clearly shows that genetic factors predispose certain people to this disease. And if one of a pair of identical twins develops MS, there is a one-in-three probability that the other twin will also develop the disease. Similarly, women whose parents had MS are 30 to 50 percent more likely than children from unaffected parents to develop the disease.

Some investigators have argued that the seemingly odd geographic distribution of MS is just the result of genetic factors, since many people of Northern European descent settled in regions above the fortieth parallel north and below the fortieth parallel south. But genetics certainly does not explain the increased incidence in northern Colorado or the observation that children under fifteen who move from low-MS areas to Colorado become just as likely as natives to develop MS. Nor do genetic theories help to explain MS clusters like the one in Galion, Ohio, or those in the Faroe Islands. These sound more like infections than bad genes. And in Galion, that infection may have risen from the dead.

We know that viruses cause some other demyelinating diseases in humans as well as other animals. In humans, several different types of demyelinating disease can arise after infections with measles, varicella, or rubella virus, or after vaccination for smallpox. Both the demyelination and the symptoms that follow in these diseases resemble those seen in MS. In addition, according to their physicians, many patients report virus-infection-like symptoms shortly before flare-ups of MS. Also, several human and animal virus infections have complex life cycles that can cause symptoms to periodically lessen and worsen. Infection with such viruses could underly "remitting/relapsing" symptoms like those seen with MS. And finally, MS patients have significantly higher levels of reactivity to several viruses. These include measles, parainfluenza, influenza, varicella, herpes simplex, human herpes virus 6, rubella, Epstein-Barr virus, mumps, respiratory syncytial virus, coronaviruses, adenoviruses, HTLV-1, and simian virus 5. We also know that the consequences of infection with any one of several viruses can vary greatly with age of the host. Varicella virus infections occur at an earlier age in temperate climates, and Epstein-Barr virus infections occur later. And Epstein-Barr virus infections that occur in early childhood are often asymptomatic, while those that occur later cause mononucleosis.

Obviously, such findings alone do not establish a clear link between a particular virus infection and MS. At the same time, these observations make it seem that a virus or other infectious agent may well be a key player in MS.

No one ever proposed that Jenny Moore might have died from an infection. And no one would have suspected that peptic ulcers or cancer or diabetes or acne might be infectious diseases. But as more diseases begin to yield their secrets to the scrutiny of modern science, more and more bacteria, viruses, fungi, and parasites turn up where they were least suspected. And more and more diseases are beginning to look like infectious diseases.

Cancer

Cancer is a complex series of diseases. Undoubtedly, many cancers involve genetic factors and inherited predispositions. And nearly all cancers also involve environmental factors—such as ultraviolet light, cigarette smoke, coal tar, and viruses. We have known for decades that some viruses cause tumors in animals. Now evidence is accumulating that viruses play key roles in the development of several human tumors.

CERVICAL CANCER

Cervical cancer ranks among the most common tumors in women. Each year, physicians in the United States diagnose about 12,800 new cases of cervical cancer. And in the United States, 4,600 women die each year from cervical cancer. About 371,000 new cases occur each year worldwide, and about 191,000 of these women die from their cancers.

Researchers have identified several factors that increase a woman's risk of developing cervical cancer: a high number of sexual partners, first intercourse at a very young age, and sex with male partners who are uncircumcised and promiscuous. All of these findings indicated to researchers that cervical cancer was a sexually transmitted disease. That, in turn, indicated that an infectious agent played a major role in the development of this disease. Today we know that certain types of human papillomaviruses (HPVs) are essential factors in the development of many cervical tumors.

Interestingly, more than 80 percent of all sexually active women will acquire HPV infections during their lifetimes. But the vast majority of these women will experience only transient infection with few or no symptoms. The few women who develop chronic infections are at a higher risk for cervical cancer. However, even among these women, only a minority will actually develop the disease. Clearly, HPV infection by itself is not enough to cause cervical cancer.

Other factors must contribute to tumor development. Likely environmental factors include smoking, large numbers of pregnancies, oral contraceptives, vitamin A and C deficiencies. And there is clear evidence that some women inherit a genetic predisposition to cervical cancer. Like the lock on an old iron safe, all the tumblers have to fall into place before the door will open and the demons can escape.

But in more than 90 percent of the women who finally do develop cervical cancer, among the demons that come together to loose the cancer there clearly is a virus.

LIVER CANCERS

Chronic infections with either hepatitis B or C virus can lead to cirrhosis and liver tumors.

According to the World Health Organization, hepatitis B virus (HBV) currently infects approximately 350 million people worldwide. Chronic infections occur in about 90 percent of infected infants, 30 percent of infected children, and only 6 percent of persons infected after age five. HBV infects approximately 1.25 million people in the United States.

Transmission of HBV requires exchange of blood or body fluids between infected and uninfected people. In infants, this happens most often via exchange of maternal blood during birth. In adults, HBV infection most often results from sexual intercourse.

Once the virus is in the blood, it infects cells of the liver called hepatocytes. The infection by itself apparently causes relatively few problems. But as the immune system moves to destroy infected and damaged cells, a major inflammation ensues. Ultimately, a large portion of the liver then becomes inflamed and hepatitis results. In adults, this inflammation usually subsides and immune responses eliminate the virus. But in about 6 percent of people, their immune systems never defeat the virus, and the HBV infection becomes chronic. This causes a prolonged inflammation of the liver, and in about one-quarter of these people, this chronic infection proceeds to cirrhosis of the liver. Cirrhosis results from death of hepatocytes and scarring of the liver. Often, cirrhosis alone leads to liver failure and death. But cirrhosis may also progress to a type of liver cancer called hepatocellular carcinoma, which is also often fatal. The final treatment for either of these diseases is liver transplantation. And, on occasion, since HBV can also infect nonliver cells, the newly transplanted liver may also become infected.

In the developed countries of the world, the incidence of HBV infection in children has decreased since the introduction and regular use of effective vaccines. But since this vaccine is unaffordable in much of the world, the current incidence of adult infections has changed little.

Hepatitis C virus (HCV) infects approximately 170 million people around the globe. About four million Americans have active HCV infections. Many of these infections result from transfer of blood from an infected person to an uninfected person. Today, that occurs most often among people sharing needles for intravenous drug use and during sexual intercourse. About three million of the four million infected Americans have developed chronic infections. And approximately 90 percent of intravenous drug users in this country have been exposed to hepatitis C virus.

Fifty-five to 85 percent of the time, infection with HCV results in chronic infection. About 70 percent of people with chronic infections will develop liver disease, and approximately 5 percent of them will die from their infections. The most common liver disease seen after HCV infection is cirrhosis. And, as with HBV, cirrhosis caused by HCV is a major risk factor for developing liver cancer.

While there is currently no vaccine that protects against HCV infection, some antiviral agents, like interferon alpha and ribavarin, do slow or eliminate HCV infections. But about 50 percent of infected men and women do not benefit from these treatments. And, again, people in many parts of the world could not afford such treatments even if they were widely available.

Of Wealth and Worms

Beyond hepatitis viruses, it is generally true that people in the developing parts of this world often have chronic infections rarely found in the wealthier places. An example is schistosomiasis. Worms, like *Schistosoma mansoni*, cause schistosomiasis. *S. mansoni* is a blood fluke. Freshwater snails release *S. mansoni* into the water. Immature forms of the worms, called cercaria, swim through the water until they attach to the skin of a human being. The soon-to-be worms then burrow through our skin and find their way to our lungs. There, they once again burrow through our cells to get into our blood, and the blood carries them to our livers. The worms work their way through livers into portal veins— the big venous systems that carry nutrients and other things from our intestines to our livers. In the portal veins, the worms lay their eggs. Our immune systems attack and, in the process, cause chronic inflammation, scarring, and portal-vein hypertension. Portal-vein hypertension causes all sorts of other problems—like varicose veins in our colons and esophagi, which, unlike varicose veins in our legs, can quickly become life threatening.

Worldwide, schistosomes chronically infect more than two hundred million people. Because of this, concurrent infections with schistosomes and HCV are common in many countries. For example, in Egypt, S. *mansoni* infects between 18 and 43 percent of men and women. Because of that, many of these people have concurrent HCV and S. *mansoni* infections. About 2 percent of Egyptians infected with HCV die from liver problems, and about 3 percent of Egyptians infected with S. *mansoni* die from liver problems. But about 48 percent of Egyptians with both HCV and S. *mansoni* infections die from liver disease. Together these two chronic infections swing a scythe much sharper than either carries alone.

Chronic infections literally open whole new cans of worms, especially when two or more microbes simultaneously infect one person. The closer we look, the larger this problem looms. Disease after disease that once seemed unrelated to viruses or bacteria, fungi or parasites, is fat with infectious microorganisms.

The World Health Organization currently estimates that 15 percent of all cancers, including half of stomach and cervical cancers and 80 percent of liver cancers, could be prevented if the relevant infectious diseases could be avoided or controlled.

Peptic Ulcers

My father frequently had problems with ulcers. The pain would keep him up at night, and it seemed like nothing he did made much difference. His doctor, an old-school Mormon bishop, told Dad to drink milkshakes to protect his stomach lining, eliminate spicy food (not a difficult thing for my father), eliminate alcohol (a minor thing for the Mormon bishop and a less-than-minor thing for my father), eat a bland diet (I could not imagine my father's diet being any blander, unless he eliminated ketchup), and reduce the stress in his life. Stress and spicy foods caused ulcers. Everyone knew that.

Well, as it turned out, stress is not the major cause of ulcers. Bacteria are.

Peptic ulcers are sores that form in the stomach or the small intestine. Human stomachs are rich with relatively nasty stuff designed to help digestion and to prevent infection. This "stuff" includes hydrochloric acid and enzymes that digest proteins—the same molecules we're made of. The only thing that stops the stomach from digesting itself is a thick layer of mucus that covers the cells of the stomach lining. That mucus shields the stomach from the acid and the enzymes.

Several things can thin or strip the mucous layer from the inside of the stomach. For example, nonsteroidal anti-inflammatory drugs (NSAIDS)—such as ibuprofen, Aleve, etc.—interfere with normal mucus production by the stomach. When the mucus is gone, the secretions of the stomach attack the lining and begin to digest it. That causes an ulcer. But aspirin and NSAIDS cause only about 10 to 20 percent of stomach ulcers and only about 2 to 5 percent of duodenal (intestinal) ulcers.

Essentially 100 percent of people with stomach ulcers and 70 percent of people with duodenal ulcers also have bacterial infections, specifically, *Helicobacter pylori* infections. And when most people with ulcers take antibiotics that kill *H. pylori*, their ulcers heal.

H. pylori has evolved a way around the acids and the enzymes of the stomach and can thrive inside of human beings. Then, when the bacterium itself or some other factor—like NSAIDS, stress, or alcohol—strips away the mucus of the stomach, the bacteria move in and attack the cells lining the stomach.

A bacterium, an infectious and transmissible biological agent, *H. pylori*, causes ulcers.

Whether we humans transmit *H. pylori* from one to another isn't clear. But recently, researchers have found loads of the bacteria in saliva from infected people. These researchers have proposed that *H. pylori* spreads through kissing, which adds a whole new perspective to romance.

But there is more to this story. About 20 percent of people under forty years of age, and about 50 percent of people over forty, carry *H. pylori* infections. The percentage of people with peptic ulcers is much, much lower. Clearly, peptic ulcers are complex. Equally clearly, peptic ulcers are an infectious disease, and in the complex production of peptic ulcers, a bacterium plays a starring role.

Heart Disease

Atherosclerosis (progressive narrowing and hardening of arteries over time) is the leading cause of heart disease in much of the world. Atherosclerosis is also the leading cause of stroke. Heart disease and strokes are the number-one and number-three killers of North Americans. Many factors appear to affect the development of atherosclerosis. The two environmental factors most often credited as leading causes of atherosclerosis are high-fat/high-cholesterol diets and cigarette smoking. However, there is accumulating evidence that bacterial infections also play a role in this disease.

Chlamydiae pnuemoniae is a bacterium that will infect many of us at some point during our lives. However, the incidence of infection appears to be highest among those of us who develop atherosclerosis. Although the connection between the bacterium and the disease remains tenuous, several factors seem to suggest a relationship. *C. pnuemoniae* does establish chronic infections that develop over long periods of time, just like atherosclerosis. And *C. pnuemoniae*, at least in experimental models, can cause persistent inflammation of blood vessels. Inflammation now appears to be a major factor in both heart disease and stroke. Though none of these conclusively links *C. pnuemoniae* to atherosclerosis, they provide strong circumstantial evidence for the importance of this bacterium.

The puzzle of heart disease remains to be solved. But more and more it seems that some of the key pieces are bacteria.

Epilepsy

Epilepsy is defined as "the tendency to have unprovoked epileptic seizures." That seems somewhat circular, but that is the nature of epilepsy. In fact, epilepsy might be more accurately described as a set of symptoms rather than a discrete disease. None of which makes much difference to a person in the rigor of a grand mal seizure, struggling to breathe and lying in a pool of his or her own urine. The only things that make any difference, then, are, Why did this happen? and How can I see to it that it never happens again?

Both are tough questions to answer. Epilepsy is the most common serious neurological disease in the world. It is likely there are many causes of epilepsy, but clearly, infectious diseases are among the leading causes, particularly in developing countries.

In the United States and other developed nations, 50 people out of every 100,000 develop epilepsy each year. In poorer countries, somewhere around 150 people per 100,000 develop epilepsy every year. And overall, about 50 million people worldwide suffer from epilepsy, with the vast majority of these living in poorer countries.

It appears that the main reason people in poorer countries are so hard hit by epilepsy is because of the higher incidence of infectious diseases in these countries.

Infections of the central nervous system are among the most common causes of epilepsy. And many of the infectious agents known to cause epilepsy are found only in poor countries. Also, infectious diseases are often more frequent and more severe in poorer countries because of inadequate sanitation and health care. All of this leads to a higher incidence of epilepsy.

Viruses, bacteria, fungi, and parasites all cause epilepsy. Many of these agents simply don't thrive in temperate climates. Because of that, much of the developed world is spared. But infectious diseases appear to be the most common preventable causes of epilepsy worldwide.

. . .

The list goes on. One study from the National Institute of Medicine alone lists nearly forty infectious bacteria, viruses, and worms associated with chronic (seemingly noninfectious) diseases in humans. None of these diseases is among those I have discussed. There can be little doubt that this list is in its infancy. Our understanding and characterization of infectious diseases and their agents have just begun.

Many diseases that we have imagined ordinary, run-of-the-mill — cancer, ulcers, autoimmune diseases, psychoses, and heart disease — are infectious diseases, caused not by our failing body parts but by little bits of other life that have crawled inside us. Our relationship with the microscopic world is more complex than we can envision. In health or disease, it is the microscopic parts of us that carry us through, or not. It is the bugs and the vermin that write the tales of our futures.

Before it killed her, multiple sclerosis turned Jenny into someone else. Most of us will never suffer like Jenny did. But none of us ever stays the same. Every day we swap a billion bits of ourselves with other human beings and we get a billion or more in return. Pieces disappear, others blossom where there was nothing. Old streams dry up. Sometimes fresh floods rise in those emptied beds. Other times, the music ends and children die.

8 ⣿

The Truth About Insanity: Infection and Behavior

The Everly Brothers—Don and Phil—gave Jim Gordon his first big break. It was the summer of 1963. John F. Kennedy was still alive. No human footprint marked the moon. James Meredith, a black man, graduated from the University of Mississippi. And Jim Gordon turned seventeen. It was a summer of promise. A time when, for many of us, the world looked hopeful.

Jim Gordon had played with a few small rock bands on and off since he was about thirteen years old. But this was different. This was the Everly Brothers—"Wake Up Little Susie," "Till I Kissed You," "All I Have to Do Is Dream," and "Bye Bye Love." This was big. When the call came, Jim could hardly contain himself. This was his chance, and he took complete advantage of it. From the moment Jim hit that stage, everyone saw that he was headed for even bigger things. His long solo riffs, his backups, his work with the sticks and the drums stood out like Jimi Hendrix must have stood out when he first played his own unforgettable version of the national anthem. Jim Gordon was good, really good.

THE TRUTH ABOUT INSANITY • 119

And Jim, with a shock of curly dark hair and a smile that lit up the stage like another spotlight, couldn't have been happier. There was nothing better than a stage full of musicians and a hall full of fans. Jim loved it. And the bands loved Jim—in the studio and onstage.

Over the next ten years, Jim played with nearly everyone who was anyone—over two hundred bands in all including John Lennon; George Harrison; Merle Haggard; Frank Zappa; Leon Russell; Jackson Browne; John Denver; Steve Winwood; Crosby, Stills, and Nash; Joan Baez; Barbra Streisand; Mel Torme; Randy Newman; John Lee Hooker; B. B. King; and many, many more.

And in 1967, Jim Gordon, along with Eric Clapton, wrote the rock classic "Layla." For that, the two of them shared a Grammy Award. To this day, Jim is a rich man because of that one song. But his money doesn't do him much good anymore. The warden won't let him have it.

In the 1970s, someone began whispering in Jim Gordon's ear. The words were spoken for Jim alone. At first, the words made little sense to Jim, but as he listened on and on, he finally recognized the voice as his mother's. Whether anyone else heard the words made no difference. Jim heard them, all the time. No one save Jim knew what his mother said to him. Whatever it was, Jim couldn't shake it, and he couldn't think about much of anything else.

Jim Gordon's musical career began to unravel. With his mother's words hammering inside his head, he couldn't concentrate, couldn't coax the rhythms out of his drums. Jim's music went to hell along with the rest of him.

People stopped calling. Jim found himself with a lot of time on his hands, and heroin stepped up to help him fill in the gaps. But even the drugs couldn't drown out his mother's voice, couldn't dam the flow inside his head.

So in 1983, Jim Gordon took matters into his own hands and went after his mother, apparently with a hammer. She died, horribly.

Schizophrenia

Schizophrenia affects more than two million Americans. It is a devastating disease. The most common symptom is hallucinations. Most often, people with schizophrenia hear voices, and the voices tell schizophrenic people horrible things. Some people see things that aren't there as well. Like John Nash—the famous mathematician portrayed in *A Beautiful Mind*—who saw people who weren't there, people who told Nash what to do and whom to fear. Sometimes, too, smells or crawling things burn inside these people. Delusions twist the world, and feelings of grandeur or paranoia often follow. Violence and aggression sometimes rear their ugly heads, and simple thoughts become torments. As in Edvard Munch's painting *The Scream*, the world warps in upon itself, and nothing is as it was.

Even if we don't personally know someone with schizophrenia, most of us have seen schizophrenics among the homeless people on our streets. Like the filthy man on the park bench behind his grocery cart with a permanent scowl darkening his unshaven face. A bottomless fear burns in the empty pools of his eyes, while he mutters or shouts to himself about men or women or beetles that aren't there as he dodges unseeable threats that materialize from thin air.

Our national policy of deinstitutionalization has put tens of thousands of mentally ill people onto our streets. Health and Human Services studies estimate that we have nearly six hundred thousand homeless men and women in the United States, and about two hundred thousand of them suffer from schizophrenia or bipolar disorder. Fully one-third of the people on our streets suffer from severe mental diseases, especially schizophrenia—condemned for a lifetime to a personal hell whose fires they cannot even name, let alone quench.

But we have no room for them in our institutions, no place but the street to care for those we deem crazy.

Despite more than a century of investigation, we still don't know what causes schizophrenia. If your identical twin has schizophrenia, then there is a 40 to 50 percent probability that you will develop schizophrenia. And if your mother or father had schizophrenia, there is a 10 percent probability that you will have the disease, too. The rest of us have only about a 1 percent probability of developing schizophrenia. That means that genes play some part in this disease. But if bad genes alone caused schizophrenia, then a genetically identical twin would always develop schizophrenia after his brother or her sister did. Something besides genes boils inside of those deranged by this disease.

Several different fingers point to an infectious agent as one of the essential ingredients for schizophrenia.

A group working at Johns Hopkins looked inside the brains of people with and without schizophrenia. Among other things, these scientists found that some genes were much more active in the diseased patients than in normal patients. Many of these genes turned out to be viral genes, in particular, genes from endogenous retroviruses. Chapter 3 described how viral genes and their fragments make up more than half of our chromosomes. Some of these genes are the burnt-out embers of ancient viral infections; others are still smoldering. All of this genetic material came from retroviruses—viruses, like HIV, that have the ability to insert their DNA into our chromosomes. Over the eons, most of these genes have fallen silent. But the group at Johns Hopkins has shown that some of these viral genes reactivate themselves during or prior to the course of schizophrenia. Activated viral genes also show up in cerebrospinal fluid from people with schizophrenia. Activation of endogenous retroviruses could be behind many of the symptoms of this terrifying disease. Then it would be pieces of our own infectious past that bubble up from the ooze to pierce our thoughts. An ancient infection risen from the rubble to twist our minds.

Other less direct evidence also points to infectious agents in schizophrenia. In 1990, the FDA approved clozapine for treatment of schizophrenia, particularly as therapy for forms of schizophrenia that were resistant to other forms of treatment. The mechanism of action of clozapine in schizophrenia is poorly understood, but the drug clearly helps some of those afflicted. We now know that clozapine inhibits the replication of retroviruses. It may be that the efficacy of this drug for treatment of schizophrenia has as much or more to do with the drug's antiviral effects as it does with its antipsychotic effects, or maybe they are one in the same.

Also, many endogenous retroviruses are most active during fetal and early postnatal life, the time when brain development moves forward rapidly. As these genes become active, they begin to replicate themselves and hop from place to place among the chromosomes. Sometimes, as the viral genes enter new chromosomes, they disrupt human genes strung out along those chromosomes. And because of their regulatory powers, once activated, endogenous retroviruses may also begin to turn human genes on or off. Any one of these events could dramatically alter the course of a developing human brain or spinal cord. And because the spinal cord and brain house big chunks of who we are and how we behave, changes in gene expression here could change a person, body and soul.

Part of the mind-ravaging disease of schizophrenia appears to be infectious. Remember, though, that infectious and contagious aren't exactly the same thing. Infectious diseases include anything caused by a prion, a bacterium, a virus, a fungus, or a parasite—diseases caused by transmissible agents. The word "contagious" is usually reserved for those diseases that can be easily passed from human to human. Creutzfeldt-Jakob disease (CJD) is an infectious disease caused by a prion. CJD is not particularly contagious. That is, you are unlikely to get it after sharing a fork with a woman who already has CJD. Influenza, on the other hand, is an infectious disease that is highly contagious.

Infections can occur in many ways, but only a few infections can be passed from person to person. So the fact that some psychoses appear to be infectious diseases does not necessarily mean that you can "catch" them from an affected person. Regardless, these findings suggest we may have to dramatically change the ways we think about "mental" diseases.

And it isn't only viruses that have been found beneath the fires that burn inside the schizophrenic. Parasites, too, may play a role.

Toxoplasma gondii is a one-celled parasite found in several mammals, including humans. But only inside of cats, most often domestic house cats, does *T. gondii* complete its life cycle and create newly infectious parasites to unleash on the rest of the world.

T. gondii infects a high proportion of people with schizophrenia. The significance of that isn't entirely clear, but it is clear that *T. gondii* infections can change the way mammals think, even the ways humans think.

T. gondii also infects rats. As I said, the parasites cannot complete their life cycle inside of rats, but *T. gondii* oocysts (an early stage in the parasite's life cycle) cannot hatch and fully develop inside of cats. For *T. gondii* to lead a normal life, the oocysts must be shed by cats and eaten by rats that, in turn, get eaten by cats.

Under normal circumstances, rats avoid cats like the plague. But rats infected with *T. gondii* develop a sudden fondness for the smell of cat urine. That leads the rats to the cats and *T. gondii* to completion of its life goals. Somehow, *T. gondii* has learned how to convince rats that they like cats. Rats made suicidal (not unlike schizophrenia) by parasitic infections: infectious insanity.

Children raised with cats develop schizophrenia about 50 percent more often than children raised without cats. And children who are breast-fed are about 1.5 times more likely than formula-fed children to develop schizophrenia. Both of which suggest a role for an infectious agent. Also, humans with schizophrenia and infected with *T. gondii* show greater cognitive impairment than people with schizophrenia alone.

Even apparently normal people infected with *T. gondii* show subtle changes in their behaviors. Nearly 50 percent of us have *T. gondii* cysts in our brains. When we garden in cyst-infected soil or inadvertently handle infected meat or change our cats' litter boxes, we can inhale *T. gondii* cysts. Inside of us, the parasite cannot pass through a complete life cycle, and clever as it is, *T. gondii* has still not evolved a plan for making humans into cat food. But that is not to say that we are unaffected by the parasite.

In tests designed to measure behavioral propensities, women with *T. gondii* cysts in their brains were more outgoing and warmhearted than uninfected women, and men infected with the parasite were more jealous and suspicious than uninfected men. Parasitic behavior modification.

Endogenous retroviruses and parasites are not much alike. So, on the surface, it is not immediately obvious why a single disease like schizophrenia might involve such different infectious agents.

Both endogenous retroviruses and *T. gondii* can alter normal brain function. But it seems improbable that these two very different agents would alter brain function in exactly the same way. Perhaps schizophrenia is not so much a disease as it is a symptom of a disease or even a series of related diseases involving alterations in brain and cord functions that result in similar constellations of symptoms.

A single disease or a series, schizophrenia—at some level—depends upon infections to alter who we are: viruses and parasites that change us, build us up or tear us down. Infections that speak to us in the voices of our mothers. A parasite that steals our thoughts. A microbe that picks up a hammer and drops it into what were once our hands.

PANDAS

From the *Journal of the American Medical Association*:

On a trip to the zoo with her family, a seven-year-old girl suddenly began to refuse to touch anything with her hands, and she demanded that the family stop at every washroom along the way so she could re-wash her hands. For the next two days, her fears worsened, "until she was unable to clean herself or use the toilet without assistance." As she washed her hands, she counted out loud to ten while she scrubbed each finger. Though she cried while she washed, she would not stop until she had finished the entire ritual. "By the time medical attention was sought, the fears and rituals had progressed to the point that her father had turned off the water to all but one of the sinks in the house and her mother had to brush her child's teeth and bathe her at least twice a day." The second-grader was also convinced that her food was unsafe and would consent to eat only a few specifically selected items.

Obsessive-compulsive disorder (OCD), with its obligatory routines, demon fears, and repetitive rituals, pulls people apart, from themselves and others. It may be hand washing and fears of contamination, like the example here from a case report in the *Journal of the American Medical Association.* Or it may be fears of infection, the imagined death of a loved one, a forgotten stove burner destroying the house, or simply complex rituals that must be performed each time a threshold is crossed or a door is opened or closed. A life unlaced by terror and mistrust.

While everyone who has ever seen them immediately recognizes the symptoms of OCD, the causes of this disease remain obscure. Obsessive-compulsive disorder is a behavioral disorder, a malfunction of the mind, a psychological disorder. Psychological disorders come from God knows where. And these diseases aren't like polio or diabetes or a lung tumor. Mental disorders are different. Aren't they?

Dr. Susan Swedo and colleagues at the Pediatrics and Developmental Branch of the National Institutes of Mental Health think not. These investigators believe OCD—especially sudden-onset, pediatric OCD—may begin just like a whole lot of other childhood diseases—with an infection.

An unexpectedly high number of children who develop OCD and other neurological disorders suffer first from bacterial infections, in particular, group A beta-hemolytic streptococcal infections, often seen as rheumatic fever or strep throat. Then a few days later, these children's behaviors suddenly take a dark turn.

As a group, these diseases are referred to as pediatric autoimmune neurological disorders associated with streptococcal infections, or PANDAS. In spite of the cuteness of the animal associated with the acronym, these diseases can destroy whole families.

First described by Dr. Swedo and her colleagues in 1998, PANDAS appear in children as abrupt-onset of OCD or tic disorders (sudden muscle contractions, often facial, that are indistinguishable from Tourette's disease) shortly after a strep infection.

There is, as yet, no obvious way streptococcal infections might cause OCD or other neurological disorders. But several different investigators have found evidence that points to a child's own immune system as the final cause of PANDAS.

Two different treatments reduce the neurological symptoms of PANDAS—complete plasma exchange and intravenous administration of mixed human immunoglobulins (antibodies). Complete plasma exchange involves swapping the affected person's plasma (the fluid portion of the blood) with plasma from an unaffected individual or individuals. The positive effect of this treatment means that something in the OCD children's plasma causes many of their symptoms. Plasma is where most antibodies are normally found. The intravenous introduction of mixed immunoglobulins probably does two things: First, these antibodies help to eliminate infectious bacteria; second, administration of large amounts of antibodies often suppresses a person's immune system.

One explanation that would fit with both of these results is a thing scientists call molecular mimicry. Some infectious agents, especially bacteria, appear to have learned how to make themselves look a little like us. They do this by subtly modifying surface molecules to make them more like ours. Sometimes, that helps these bacteria avoid destruction by our immune systems. Other times it backfires on everyone involved.

If the bacteria's efforts at mimicry aren't perfect, a human's immune system will still recognize and respond to the bacteria—which is good. But because parts of the bacteria look nearly human, as the immune system responds to the infection it also may begin to respond to self. When that happens, autoimmunity follows—which is bad.

If PANDAS truly are the result of molecular mimicry, the smoking gun would be finding a child's own antibodies bound to something, like his or her brain, that could change the way the child behaved. Several laboratories have now found such antibodies. And there is no doubt that these antibodies, which bind to the basal ganglia in children's brains, could cause the kind of changes seen with OCD.

Then this would be a behavioral disorder with an infectious root. Maybe mental diseases aren't so different from mumps and measles and tumors after all.

Crazy Animals

We don't normally think of animals as going crazy. "Crazy" is a term we reserve for people, because we believe that mental diseases are unique to humans. We have minds that we might lose, but others don't. The downside to this point of view is that it can limit how we think about, treat, and deal with mental diseases. For lack of that, some people suffer.

Animals besides humans clearly do go crazy. What causes this animal craziness is surprising.

Earlier in this chapter we saw how a parasite (*T. gondii*) can make rats fond of cat urine and, as a side effect, suicidal. We don't call them crazy rats or schizophrenic rats or deranged rats. Scientists call them infected rats instead. That is because, with rats, our own images of ourselves don't get in our way. With rats, we understand that behavior — just like digestion or breathing or walking — depends on everything else, especially everything else infectious. But rats infected with *T. gondii* are clearly as crazy as any human we have ever called crazy. These animals lose all sense of self-protection and lose the instinct and drive for survival. That's crazy.

And it isn't just rats that are driven mad by infections.

Ants infected with parasites or molds also turn suicidal, because just as with *T. gondii*, the ants' suicides benefit the infectious microorganisms.

Dicrocelium dendriticum is a liver fluke, a flatworm that infects cattle. This worm has one of the more complex life cycles known. Life, for this worm, begins in cattle. Inside the bile ducts of cattle, *D. dendriticum* lays its eggs. Then, along with the bile, the eggs pass out with the feces onto the grasses of the fields where the cattle graze. Among those grasses are land snails that feed on cattle dung. As the snails consume the dung, they also consume the eggs of D. dendriticum. Inside of the snails, the eggs hatch and become cercaria, an early stage in their life cycles that begins these worms' complex lives. The cercaria pass out through the feet of the snails as the snails secrete slime balls to aid their progress across the fields. Also scattered among the grasses are ants that feed on slime balls. As the ants eat the slime balls, the ants ingest the cercaria. Inside the ants, the worms mature into metacercaria.

There, if it weren't for the parasite's genius, everything would come to a screeching halt. Normally, these ants carry out all their affairs very near the ground. By doing so, the ants avoid being eaten along with the surrounding grasses as the cattle graze. D. dendriticum-infected ants appear perfectly normal at first. But as soon as the sun goes down and the temperature drops, the infected ants go crazy. Instead of maintaining their low-profile existence, the infected ants climb to the tops of the grass spears, clamp in their mandibles, and practically beg the cattle to eat them. If infected ants survive the night, when dawn breaks, they return to normal and go on about their affairs as though nothing had happened. But as darkness comes again, once more the madness takes them.

We don't normally call ants crazy either. But the behavioral disorder among D. dendriticum–infected ants is as severe as any seen in humans with diseases like schizophrenia, depression, or bipolar disorder. These ants are disturbed, severely disturbed.

D. dendriticum is most common throughout much of Europe and Asia, but D. dendriticum is also found in parts of North America and Australia. The most common mammalian hosts for D. dendriticum are pigs, sheep, and cattle. But D. dendriticum does infect humans, mostly in Africa and China. The consequences of that have never been fully investigated.

In South America, ants infected by a mold called Cordiceps also lose their sensible fear of heights and climb to the tops of long spears of grass. There, the mold sprouts through the ants' skulls and the wind carries off the mold spores. If the ants remained sane and on the ground, the spores of *Cordiceps* would never spread as widely as they do. Severe behavioral modification that comes not from years of therapy or poor parenting but from infection.

And there is more.

Shrimp and fish infected with another flatworm also behave dangerously. Normally, both the fish and the shrimp live and feed at deeper ocean levels where the water above protects them from predatory birds. But infected fish rise to the surface and often turn on their sides, presenting bright, shallow targets to gulls and others. These fish are thirty times more likely than uninfected fish to be eaten by birds. Infected shrimp also behave oddly. They rise to the surface of the water, swim erratically, and again quickly attract the attention of hungry birds.

Inside the birds, the worms complete their life cycles and then return to the waters below—seeking once more to drive fish and shrimp mad.

Herpes viruses sometimes travel up neurons into people's brains, where the virus can cause encephalitis. Dramatic changes in behaviors often follow. HIV, by other routes, can also reach into people's brains and change the ways people see the world and those around them. As we saw in Chapter 1, *Treponema palidum*, the bacterium that causes syphilis, can move from a person's genitals into his or her brain and heighten his or her desire to mate, then alter everything about a person—gait, vision, mind, future.

And, of course, there are the transmissible spongiform encephalopathies like kuru, mad cow, Creutzfeldt-Jakob disease, and its variant. All these are caused by prions—bits of protein that enter our mouths and crawl into our brains where, piece by piece, these proteins dismantle human beings.

Infections change who ants and fish and shrimp and rats are. And infections change who we are. Even among those of us we might call normal, infection changes us. Chapter 2 explained the weird and unexpected fates of sterile animals: Normal digestion, immunity, development, and behavior all depend on infection. Perhaps what we call normal or abnormal human behavior is nothing more than a reflection of what is or isn't infecting us. Maybe the variety of human behavior is only a reflection of the mixtures of microbes that we carry with us.

David Vetter, the boy in the bubble, was the only human I know of who was never infected. David Vetter was certainly not normal. No doubt his years of isolation changed him. Living inside a plastic bubble and never being allowed to touch another human being would change anyone. But maybe there was more. Maybe our infections keep us sane as easily as they drive us crazy. Maybe a piece of David was missing, a piece no human can do without.

The rest of us live with our infections. We don't imagine that those infections change us much. Our bacteria, our viruses, our fungi, our parasites are just passing through.

But our imaginings don't fit very well with the data. We are complex beings with too many parts to number. Which of those parts make us who we are? Which are too small to take seriously? Too chemical? Too microbiological? Too terrifying?

The judge and the jury understood full well that Jim Gordon was crazy. And whether it was an infection, a mental aberration, poor toilet training, or something else made no difference; his sickness was dramatic and he needed help. Years before, the State of California had decided that regardless of what made people like Jim sick, its good citizens didn't want to run into any of these people in some out-of-the-way place on one of their bad days. So, California law now demanded that all such offenders be locked up. Because of that, neither the judge nor the jury had a choice. That day in 1984, the court found Jim Gordon guilty of second-degree murder, and the judge sentenced him to sixteen years to life in prison. And behind all that concrete, chain link, razor wire, and iron bars, Jim's mother just went on whispering into his ear.

9 ⠂⠒⠂

Red Dawn:
The Shape of Things to Come

It's hot. Even for Minnesota in August, it's hot. Jimmy Wilson squats at the weedy bank of a small pond, his knees caked in mud. He eyes himself in the still reflection of the dark water. Blond-haired and blue-eyed, Jimmy is handsome and large for his eleven years. The wind tosses his hair, and Jimmy admires the way it falls across his face. The air is rich with the scents of mud and rotting greenery and the thick, sweet smell of stagnant water. The other children are on the opposite side of the pond now, out of sight. Jimmy's teacher stands in between, watching them all and pleased with herself for having brought the children here. So much to see and smell and touch and hear, so natural, so environmental.

As Jimmy continues to admire his image in the pool, he notices another pair of eyes staring back at him with an unblinking, flat stare. *It looks like a frog*, Jimmy thinks, and he reaches for it. His right hand slides into the tepid water and around the familiar shape and slime of a leopard frog. Excitedly, he pulls what he's caught from the pond.

But as soon as he lifts the animal from the water, he tosses it into the grasses at his feet and staggers backward. Stunned, the animal just lies where Jimmy has thrown it. Jimmy drags his dry hand across his face, trying to think, trying to get hold of this place once more.

Get Ms. Hawthorne, he thinks. *Go get Ms. Hawthorne.*

He begins to run.

"Ms. Hawthorne, Ms. Hawthorne. There's something terrible here! Terrible."

"Jimmy," she calls to him as she grabs hold of his hand, "calm down, settle down, catch your breath." And she holds the child still against her brown arm.

Jimmy begins to breathe again.

"Now tell me," she says. The young woman with close-cut hair and dark eyes reaches to brush the hair from Jimmy's eyes, startled by his fear.

"If you don't believe me, you can see for yourself."

"Jimmy, what is it?"

"Something terrible."

"What?"

"Some kind of animal."

"Jimmy, I'm sure it's nothing. Maybe it was just a shadow or an old log. Let's go take a look."

"It wasn't a log, Ms. Hawthorne. It wasn't anything like a log."

"Let's see, Jimmy."

Ms. Hawthorne, Jimmy, and some other boys and girls who saw what happened follow Jimmy back to the opposite edge of the pond. For a minute or two, Jimmy can find nothing. He begins to worry that whatever it was, it has crawled off and no one will ever believe him about what he saw.

But there it is, still lying where he flung it. Still terrible.

"Here!" he shouts.

Everyone moves to the spot he is pointing at, and they all look carefully into the low weeds.

It looks like a frog, at first. It is the right size and shape and color. But it isn't quite a frog either. There are two round flat frog eyes staring back at the children, but the eyes are staring at them straight out of the frog's back.

Ms. Hawthorne bends to take a closer look. While she studies the not-quite-a-frog, a few of the children move off to look at the pond where Jimmy found this thing.

The woman stares at the frog. It begins to stir in the grass. *It is a frog,* she thinks. And she is enough of a biologist to know that sometimes things like this—mutants—happen. But still, the creature unsettles her. The eyes blink at her.

Abruptly, the children are back from the pond. One carries a frog with a pair of legs growing out of its chest and another child has a frog with only two toes on each foot. Someone else holds up a frog with no legs, and then one with three hind legs. One little girl cups a frog with no front legs, and a young boy carries a frog without eyes.

The meadow, the ponds, the murky blue sky all begin to swirl around Ms. Hawthorne. She braces herself against the ground and tries to collect herself. She breathes very deeply.

"Leave them. Scott, Melanie. Leave them. And you, too, Ellen. All of you, leave the frogs here. We need to get back to school now."

The Future

The names are fictional. But the event happened more or less as described. The mutant frogs are real, and they are turning up in more and more places every day.

I've already talked about the many ways bacteria have worked to shape this planet and about how many bacteria there are in this world. It is unimaginable that any macroscopic change in this world wouldn't cause microscopic changes. And while we may understand that what human beings are doing to this world is changing it, we have overlooked the most immediate and perhaps the most severe consequences of those changes. Movies, like *The Day After Tomorrow,* have warned of the dangers of global warming and its effect on the polar ice cap. And movies like *China Syndrome* predicted nuclear disasters like the one at Chernobyl. But it seems no one imagined that the first consequence of our deeds might be simply to move bacteria or viruses or parasites from one place to another. No one ever expected that simple alterations to a perilous balance that took us four billion years to establish—the balance between us very dispensable large carnivores and the invulnerable microscopic world—could cause disastrous change.

From the 2000 Central Intelligence Agency report, *The Global Infectious Disease Threat and Its Implications for the United States:*

- Twenty well-known diseases—including tuberculosis (TB), malaria, and cholera—have reemerged or spread geographically since 1973, often in more virulent and drug-resistant forms.
- At least 30 previously unknown disease agents have been identified since 1973, including HIV, Ebola, hepatitis C, and Nipah virus, for which no cures are available.
- Of the seven biggest killers worldwide, TB, malaria, hepatitis, and, in particular, HIV/AIDS continue to surge, with HIV/AIDS and TB likely to account for the overwhelming majority of deaths from infectious diseases in developing countries by 2020. Acute lower respiratory infections—including pneumonia and influenza—as well as diarrheal diseases and measles, appear to have peaked at high incidence levels.

The courses of our lives are changing. Once, we might have thought that only bombs or missiles could force great changes. Very few of us would have anticipated that the little everyday things that humans do would have the greatest impacts on our futures. Nor would we have expected that the first signs of those impacts would be microscopic change. But that is how it appears to be—microscopic changes are eroding our macroscopic world.

And those few of us insightful enough to have seen some of this coming might have predicted that the effects would be felt mostly in the developing countries where there are the fewest controls over human activities.

But we would have been wrong.

Impact Within the United States

Because of its wealth, the United States is relatively insulated from most of the changes that occur in this world. But even in the United States, deaths from infectious diseases nearly doubled between 1980 and 2000 to more than 170,000. Some are quick to point out, though, that most of the agents that have caused major outbreaks of infectious diseases in the United States, like West Nile virus, originated in other countries. But in this mobile world, that claim loses much of its importance. Regardless, it is possible to make certain predictions about the role of infectious diseases in the future of the United States. These come from a publication by the U.S. National Academies: "In the opinion of the US Institute of Medicine, the next major infectious disease threat to the United States may be, like HIV, a previously unrecognized pathogen. Barring that, the most dangerous known infectious diseases likely to threaten the United States over the next two decades will be HIV/AIDS, hepatitis C, TB, and new, more lethal variants of influenza. Hospital-acquired infections and foodborne illnesses also will pose a threat."

HIV/AIDS deaths continue at a nearly constant rate in the United States. The likelihood of an effective vaccine remains remote. Add to that the inevitability of continued evolution of the virus, and it seems highly likely that HIV/AIDS will continue to kill North Americans in record numbers for the foreseeable future.

Hepatitis C currently infects more than four million people in the United States, and the numbers are climbing rapidly. There is no effective vaccine against this infection and, because of that, some predict that within five years, deaths from hepatitis C infections will outnumber deaths from HIV/AIDS.

Between 1953 and 2004, the number of deaths from tuberculosis in the United States dropped from roughly eighty-four thousand to about fourteen thousand. Even so, with the appearance of several new antibiotic-resistant strains of tuberculosis, the spread of HIV infections and the associated immune suppression, and immigration—especially illegal immigration—it is highly likely that there will be a resurgence of tuberculosis and related deaths in the United States. Because illegal immigrants are not subject to any medical screening before entry into the United States, their contribution may be larger.

As of now, about thirty thousand people in the United States die from influenza each year. It is hard to predict when the next pandemic may hit. On the other hand, it is very easy to predict whether another pandemic will hit—certainly. Influenza is discussed in much greater detail in Chapter 15. At this point, it is sufficient to say that influenza deaths in the United States will vary in unpredictable ways—some years will undoubtedly see fewer than average deaths, some years will see more, a lot more.

Hospital-acquired infections already kill more than fourteen thousand people a year in the United States. The continuing evolution of several virulent microorganisms will unquestionably produce new antibiotic-resistant strains, as has already happened with staph. That ensures that the number of deaths from infections with these bacteria will rise, though it is impossible to predict when and just how steeply the numbers will rise.

Between 1995 and 2000, the United States doubled the quantity of food that it was importing from other nations. The tons of food we import are almost guaranteed to continue to increase. Currently tens of millions of people are sickened each year by foodborne microorganisms, and approximately ten thousand die from these infections. Because of the mobility of U.S. citizens and the country's growing reliance on the products of other countries, it is no longer relevant whether a disease originated in the United States or elsewhere. The world's problems are the United States' problems.

And the world's problems are considerable.

Sub-Saharan Africa already accounts for more than half of the world's deaths from infectious disease. Deaths from infectious diseases in these countries will clearly continue to increase, due to newly emerging and reemerging disease and, especially, HIV/AIDS-related deaths. Coincidentally, Africa's health-care crisis will worsen as national budgets are stretched even further, political instability spreads, and personal income continues its downward spiral.

In Asia, multidrug-resistant tuberculosis, malaria, and cholera are continuing crises. At the same time, HIV/AIDS cases are increasing dramatically. In fact, some estimates suggest that by 2010, the number of deaths from HIV/AIDS in Asia may equal or surpass those in Africa. In addition, most of the worldwide deaths from H5N1 avian influenza among animals and humans have occurred in Asia.

Among the states of the former Soviet Union, the incidences of diphtheria, dysentery, cholera, and hepatitis B and C are all on the rise. And these states, including Russia, have a raging epidemic of sexually transmitted infectious diseases, especially HIV/AIDS. In Moscow alone, more than 30 percent of the homeless carry at least one sexually transmitted infectious disease. In a recent study of three detention centers for homeless people, 32 percent of the people were infected with syphilis, 17 percent with gonorrhea, and 11 percent with chlamydia (another sexually transmitted infection). Nearly a million Russians are currently infected with HIV, and that number is expected to rise dramatically in the next few years.

In Latin America, great progress has been made in elimination of some infectious diseases, such as polio, but much of this region remains desperately poor. Because of that, cholera, malaria, tuberculosis, and dengue fever (see Chapter 11) are all on the rise.

The Middle East and North Africa have ongoing epidemics of tuberculosis, as well as hepatitis B and C. However, this region does benefit from considerable wealth and stringent moral codes that apparently have limited the spread of HIV and other sexually transmitted diseases. Currently, this region has one of the lowest reported HIV-positive indices in the world. Some, however, have questioned whether this is truly due to a very low rate of infection or to severe underreporting of infections because of the social stigma linked to the disease.

Conditions in Western Europe are more like those in the United States. Though there are ongoing problems with HIV/AIDS, hepatitis B and C, and tuberculosis, the relative wealth of this region coupled with the excellent standard of available health care suggest that infectious diseases will have less of an impact here. However, the high level of tourism and travel into and out of this region of the world put it at a high risk for the importation of infectious diseases. In addition, as of now, the avian flu pandemic that hit Asia and Eastern Europe is spreading into Western Europe.

Encounters with prophets open holes in people. Elijah's tales of the sinister Ahab fall on Queequeg and Ishmael's deaf ears, because the old sailor tries to tell them something they don't want to hear. Oedipus ignores the Oracle at Delphi—but as foretold, the boy murders his father and takes his own mother as wife. Cassandra's warnings fall on her mother's uninterested ears. Queequeg pays for his blindness with his life. Oedipus plucks out his eyes and his mother kills herself. Cassandra is imprisoned and Troy destroyed. Ignoring prophets can have dire consequences.

That puts the rest of us between a rock and a hard place. On the one hand, we know that many predictions don't ever come true, and mostly we prefer it that way. On the other hand, we know that some predictions come to pass. And with predictions about infectious diseases, the truth (if that's what it is) in these predictions is too horrifying to ignore.

As Chapter 6 discussed, between 1973 and 1998, twenty familiar diseases—like cholera, malaria, and tuberculosis—reemerged or spread dramatically. During the same time frame, at least thirty new diseases appeared around the globe. Since the World Health Organization published those data, seventeen more diseases have either first appeared or reappeared in the world.

Severe acute respiratory syndrome (SARS) emerged in the Guangdong Province of southern China and spread to thirty countries and administrative regions within six months. Hendra virus appeared in 1994 in Australia and again in 1999 in Malaysia. Half of all the cases so far have been fatal. Enterovirus 71 (EV-71) caused a major epidemic in Taiwan in 1998 and again in 2000. Enterovirus 71 is one of the causes of epidemic hand, foot, and mouth disease (HFMD). There have also been EV-71 outbreaks in the United States and Bulgaria. In June of 2003, monkeypox infections flared in Wisconsin, and by the end of that month, the virus had spread to Illinois, Indiana, Kansas, Missouri, and Ohio. West Nile virus appeared in the eastern United States in 1999, and by 2003, nearly every state in the union reported illness and deaths due to that virus. In 2000, several people in California died after being infected by Whitewater Arroyo virus, a member of the same family of viruses that includes Lassa fever virus and lymphocytic choriomeningitis virus. In June of 2004, cyclosporiasis, a sometimes fatal parasitic infection, erupted in Pennsylvania after several people ate raw Guatemalan snow peas. And as of April 2005, an outbreak of Marburg hemorrhagic fever (caused by a virus closely related to Ebola virus) in Angola had infected 205 people and killed 180. In spite of intense efforts to contain this outbreak, it continues to spread. Meanwhile, another H5N1 strain of influenza arose in Southeast Asia, and Rift Valley fever, plague, Ebola hemorrhagic fever, dengue virus, cryptosporidiosis, diphtheria, hepatitis C, Lassa fever, cholera, yellow fever, and HIV/AIDS spread and reemerged as major infectious-disease threats. Antibiotic-resistant strains of old scourges like *Staphylococcus aureus*, malaria, and tuberculosis also appeared.

Lately, microscopic life has outstripped even the most imaginative predictions.

In her book *The Coming Plague*, Laurie Garrett offers some insight into this. She describes the example of Lyme disease and the consequences of changes in the numbers of deer and large predators. If it hadn't been for AIDS, Lyme disease would likely have been seen as the major scourge at the end of the twentieth century. Virologists named the disease after Lyme, Connecticut, the town where it first appeared. Since the disease's discovery in the 1970s, the infection has spread to all fifty states and parts of Europe. A bacterium, *Borrelia burgdorferi*, causes Lyme disease. Ticks transmit the disease primarily from deer to humans. Once infected, people develop skin rashes, severe arthritis, and sometimes encephalopathies that cause loss of memory and cognitive function. It's a serious illness. The ticks that transmit *Borrelia burgdorferi* feed primarily on deer. Historically, the numbers of these deer were relatively low and their distribution limited. But humans destroyed most of the large predators—wolves, bears, mountain lions. As a result, the deer population ballooned. At the same time, human construction was encroaching on traditional deer habitat. Oddly enough, reforestation of land close to homes may also have been a factor. Out of the woods came deer infected with a venomous spirochete. Abruptly, people became food for the deers' ticks. And another epidemic was under way. In this case, most of it happened simply because we humans hadn't imagined what microbial consequences might accompany the destruction of one or two large animal species.

Environmental Change and Human Disease

From the United Nations report *Environmental Changes Are Spreading Infectious Diseases*:

> 22 February 2005—Dramatic environmental changes now sweeping the planet, such as the loss of forests and the spread of cities, are promoting conditions for a rise in new and previously suppressed infectious diseases, including malaria and bilharzias (also known as schistosomiasis), according to the United Nations Environment Programme (UNEP) latest yearbook.
>
> Noting that one of the Millennium Developments Goals (MDGs) adopted in 2000 by the UN Millennium Summit seeks to reverse the spread of HIV/AIDS, malaria and other diseases, UNEP Executive Director Klaus Toepfer said: "If environmental degradation is not checked then, it is clear from these new findings that this will be harder and tougher to achieve."

I mentioned the role of environmental changes in the emergence of Lyme disease in humans. But Lyme disease is only one example among many.

Malaria is one of the leading causes of sickness and death worldwide. A single-celled parasite causes the disease, and mosquitoes transmit the parasite between human beings. Mosquitoes breed best in warm, moist climates and especially well in warm, standing pools.

Mining operations and widespread deforestation have dramatically expanded such environments for breeding, especially in Africa. Deforestation has increased mosquito habitat by stripping trees and increasing erosion. And mining operations have created both tailing ponds where more mosquitoes breed and increased human susceptibility as a result of mercury pollution and immune suppression.

And the mosquitoes are moving, especially into what were once more temperate zones. Global warming—the change of average temperatures by even a few degrees—is unlocking the gates that once held the mosquitoes near the equator. WHO now estimates that 6 percent of all human cases of malaria result from environmental changes.

Somewhere between one million and three million people die every year from malaria, 75 percent of them African children. Six percent of this number is 60,000 to 180,000 human lives each year, 75 percent of them children. And the mosquitoes' habitats are expanding.

A parasitic worm causes schistosomiasis. This disease often results in lifelong, painful, sometimes paralyzing, infections. As discussed earlier, people acquire this disease by wading or swimming in water where worm-infested freshwater snails live. The snails release the parasites into the surrounding water, where they can live for up to forty-eight hours. If during those forty-eight hours a swimmer or bather enters the pool, these parasites can burrow through human skin and set up shop inside that man or woman.

Like the mosquitoes that cause malaria, the snails that carry schistosomes are spreading. Fewer trees, more water, and warmer climates all favor the overproduction of the snails, mosquitoes, and the spread of the disease.

. . .

The National Institute of Allergy and Infectious Diseases in Washington, DC, has also issued warnings about the potential impacts of environmental changes, some of which, though apparently "natural," may ultimately result from human activity.

Valley fever is endemic in the southwestern United States. A fungus called *Coccidioides immitis* causes valley fever. Heavy rains following years of drought seem to help raise the fungus from the soil. High winds then spread it to human noses, where it can cause serious disease. Changes in annual rainfall clearly happen spontaneously, but some may also follow changes in human activity, such as drought as a consequence of global warming. But truly "natural" changes in the environment can also create infectious problems. For example, the 1994 Northridge earthquake may have contributed to the public health problem from valley fever by ejecting fungal spores into the air.

In 1993, there was a sudden upsurge in the incidence of hantavirus (which causes an often fatal respiratory disease) infections in the four-corners area of the United States—the spot where Utah, Arizona, New Mexico, and Colorado come together. The evidence now suggests that the major factor in this outbreak was an unusually rainy spring that led to an abundance of vegetation which, in turn, led to an abundance of the deer mice that carry hantavirus.

Relocation of rivers, construction of dams, and drainage of swamps all appear to alter the distributions of infectious agents and their vectors. Following the construction of a dam in the Senegal River basin in 1998, in western Africa, there was a major outbreak of Rift Valley fever (a disease caused by a virus transmitted by mosquitoes). And in Africa and the Middle East, dam construction led to outbreaks of schistosomiasis.

Deforestation is occurring at an alarming rate throughout the world and the subsequent settling of these newly cleared areas appears to be responsible for outbreaks of a host of viral, parasitic, and bacterial diseases. In Brazil alone, the expansion of gold-mining operations into old forests appears to have resulted in a significant increase in the number of cases of malaria.

The reverse may be true for Lyme disease in the United States, where suburban growth, predator control, and some reforestation have increased tick populations and the incidence of the disease.

The largest numbers of living things on this planet are microscopic. Nothing happens here that bacteria and viruses and fungi and parasites don't notice.

Microbial Change and Human Disease

When we push them, they push back. As described in Chapter 6, living creatures evolve, and bacteria, viruses, fungi, and parasites evolve much faster than we do. We bury them under a pile of antibiotics, and like the phoenix, they rise from the ashes, newly resistant to everything we have thrown at them, and attack us once more.

Or we take the grizzlies, the wolves, and the lions out of our lives, and the plague of deer that follows brings with it a whole new set of human diseases. Or we build a dam and open a new world to mosquitoes. Or somewhere in the tropics an old virus grabs hold of a new gene and makes the leap from chimp to human, and our world shifts.

We have always believed that certain species of animals, usually vertebrate animals like frogs or birds or gorillas, were among the most sensitive indicators of environmental damage. Maybe not. Maybe the most sensitive indicators are changes in the patterns of infectious diseases. After all, nothing on this planet, including the planet itself, is uninfected. Nothing is as thoroughly interwoven with all life on this planet as microorganisms. Perhaps that is where we should look first for images of our future.

A Plague of Mutant Frogs and
Other Omens

After Ms. Hawthorne gathered her students and stumbled back to school, she immediately called the Minnesota Division of Wildlife. Though shaken by her experience, she made it very clear what she and her class had seen that August day in 1995. In the process she opened up a very large can of worms.

By 2001, the Northern Prairie Wildlife Center had received nearly two thousand reports of malformed frogs and verified 835 of those reports. Fifty-four different species of amphibians in forty-four states and four Canadian provinces were affected. At that point, it appeared that about 10 percent of the total frog population and many other amphibians were exhibiting similar deformities.

Something was mutilating the United States and Canadian frog populations. What was it?

At first, most everyone suspected environmental pollutants. Frogs, especially as tadpoles, are thin-skinned and absorb many chemicals directly from the water in which they swim. That makes frogs especially susceptible to toxic changes in the water. Second, people suspected that increases in ultraviolet light, due to a thinning ozone layer, might be responsible. And it still seems that pollutants and ultraviolet light have played a role.

But, as of now, most of the malformations appear to be the work of a parasitic worm. That is, an infectious disease is causing most of the horror racing through North America's frogs.

This particular parasite is a trematode or flatworm called *Ribeiroia ondatrae*. These worms burrow into developing tadpoles and alter the normal processes that generate frogs from tadpoles. When the parasite's work is done, frogs appear with a tangle of a dozen limbs, no limbs at all, limbs in the wrong places, or eyes in the backs of their heads.

R. ondatrae has actually been torturing frogs for fifty years or more. With that discovery, everyone heaved a great sigh of relief. It was just a trematode after all, an infectious disease, admittedly, but nothing we created. And it had been going on for years.

But when people took a closer look, they realized that while this might have been going on for some time, no one had ever reported the numbers of mutant frogs that we are seeing today. Something shifted the balance between frogs and worms. It is likely that we humans were the motivating force behind that shift.

And it isn't just frogs and other amphibians. Infections in seals and dolphins have caused massive beachings of these marine mammals throughout the world. The cause appears to be a morbillivirus, similar to the virus that causes measles in people, distemper in dogs, and rinderpest—a highly fatal disease in cattle. It appears that PCBs that leaked into both fresh and marine waters have depressed these mammals' immune systems and opened the doors for the morbillivirus.

Chronic wasting disease, a disease much like mad cow disease, is spreading through elk, deer, and now moose herds in the central United States and Canada. The disease, which causes severe wasting and staggering in elk and deer, apparently began in Colorado. From there it spread to wild and ranched populations of deer and elk in Wyoming, Nebraska, South Dakota, New Mexico, Utah, Wisconsin, Illinois, and Saskatchewan.

The roots and the routes of this epidemic of chronic wasting disease aren't understood. But at the same time that this disease is spreading, we are bulldozing elk and deer habitat. Many of the large carnivores that feed on deer and elk have dwindled or disappeared. And because of drained wetlands and fouled streams, the quality of the water and the forage available to these animals has deteriorated as well. While no one factor clearly contributes to the spread of chronic wasting disease, combinations of these factors have certainly pushed other diseases into new territories.

We don't know exactly which of the human-caused changes is most important to the spread of infectious diseases. Something we've done, though, appears to have helped with the spread of trematodes among amphibians—chemical pollutants, increases in ultraviolet light, over-

growth of algae because of changes in water composition are all possibilities.

Crystal Gazing

In August of 1995, in south-central Minnesota, a group of schoolchil-
dren from Minnesota New Country School in Le Sueur set out on a
field trip to study the ponds and bogs of the region and to collect
samples. What those children found that day would shake the entire
country.

We push, the microbes push back. Every time we pluck a single
strand, the whole web vibrates. No one can predict what forces those vi-
brations will rouse.

But even if we didn't touch a thing, the world would change. And
any change that flows through this world changes the microscopic as
surely as the macroscopic. Cosmic radiation, ultraviolet radiation, sex,
deletions, inversion, infections, and transpositions are at work all of the
time. Moving genes, whole pieces of chromosomes, sometimes even
whole genomes between individuals change the rules of the game even
as the players are sitting down to the table.

Still, it requires no crystal ball to see the outlines of our futures. The
emergence of new and revitalized old infectious agents is inevitable.
The places, and the bugs, and the number of dead are hazy. The rest is
clear, though, clear as a Minnesota pond and the pale yellow eyes of
feral frogs.

part three

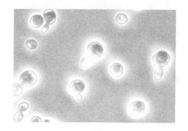

MICROBES THAT WILL CHANGE THE WORLD

10 ⣞

The Spider in Room 911: SARS

Saturday, November 16, 2002. Foshan,
Guangdong Province, People's Republic of China

Clinging to the handrail, the old man eases his way onto the small wooden platform. *If I just sit for a moment,* he thinks to himself, *I will feel better. Then I can go on to my office and do my work.* He pulls his cap from his head and collapses onto one of the benches.

Over his head, the tiled roof of the small park bench arches into traditional Chinese wings that reach for the blue sky. Back bent, arms propped on his knees, the old man stares into the pond that spreads out in front of him. Foshan, this town, his town, is like a butterfly to the great tigers of Beijing and Shangai. He has lived in this town all of his life. All of his life, this town has been enough for him. He is stooped now with his illness—some flu or pneumonia or . . . he cannot put words to it, and Foshan seems far away from him. The sickness came on him suddenly. Only two days ago, he had felt well—mostly. Now he feels he cannot walk another step.

His gray head hangs forward, and the chill of the November air works at his scalp. He coughs a racking wheeze that exhausts him. He lifts a soiled handkerchief to his lips. A little mucus darkens the cloth as he draws it away and stares at the gray spot.

For years he has sold his goods to the many cities and shops of Guangdong Province in southern China. Guangdong is large—more than seventy-five million people—but he has always come home to Foshan. He is a man others take seriously here—a thing he was very proud of, until yesterday. But now none of it makes much difference. He leans farther forward, holding his stomach tightly, and he is taken by the twisting grip of another series of coughs. The dark eyes behind his thin-rimmed glasses falter. He swipes at his mouth once more with the cloth.

A little more rest, he thinks. *Just a little more*. Then he closes his eyes and lays his head back against the worn rail of the wooden bench.

The few humans who wander the cold park on this particular morning take no notice of the old man's death. And it will be hours before anyone thinks to ask the man if he is all right. He won't answer, of course. Then, uniformed men will come and move the linen-white body out of the park and deliver it to the hospital for autopsy.

A small report will be filed listing "atypical" pneumonia as the cause of this death. But he was an old man, so no one will give too much thought to his dying or wonder how many others' lives he might have touched before he died.

A day or so later, everything in the little park in Foshan seems the same as it was before the small man came here to die. But it isn't. With the old man's death, the world has shuddered. It will be another month or two before anyone notices that. Then all hell will break loose, and at the center of the storm will lie the old man's body.

Week 1. Thursday, February 20, 2003.
Hong Kong, Special Administrative Region, China

Near the bay, a physician strolls with his sister's husband along the busy streets of Kowloon. As they walk, the younger man, the brother-in-law, points to the old buildings and the ships in the bay, showing his friend the sights. The two men speak idly of their pasts together and their plans for the future. Though there are ten years between them, they have much in common. Today, they have wandered together for hours, the words spinning out before them and the day spinning out from underneath them. Both are a little gray at the temples, a little bent in the shoulders. They have dressed lightly, for the winter here is broken by the heat of the sea. Toward evening, the two sit together in a small bar and share a bottle of wine. The men eye the smartly dressed women who pass. The evening air is a melange of hot oils and fresh fish, diesel and jasmine, and the womanly smell of the sea. Finally, it is time to say good night. The two men shake hands warmly and embrace. Promises are made about meeting again soon, cards exchanged. More promises. Then one heads off for his home in the country and the other makes his way along Waterloo Road to the Metropole Hotel.

The physician has traveled all the way from Zhanjiang at the southernmost tip of Guangdong Province. For a few days now, he has been coughing and running a fever that he cannot seem to shake; there has been some diarrhea, too. But he came here to visit family, and he is convinced the illness will pass. For now, it is not so bad. Today he felt well enough to sightsee with his brother-in-law, and he enjoyed it.

He glances down at his watch. *Good Lord,* he thinks, *we spent more than ten hours together today. And already I am looking forward to the next time, whenever that may be.* The man smiles at the prospect and walks on toward his hotel. His room is on the ninth floor, and he likes the view from there.

But there will be no next time for this man and his relatives. The thing that killed the old man in the park in Foshan will see to that, and it will see to it very shortly.

By morning, the physician's "flu" has worsened severely. His temperature is up, he is dehydrated from a night of diarrhea, dead tired, and his cough feels as though it is pulling his lungs out. With the little energy he has left, he rises, bathes, and dresses himself—tan slacks, blue shirt, a tie. Then he calls down and asks the concierge to call a cab.

Once settled into the taxi, he asks the driver to take him to the nearest hospital. The physician from Guangdong is a nephrologist, so he knows enough medicine to realize that his pulmonary problems are serious.

In the emergency room, the doctors move quickly to assess the man's condition. A nurse, working on another patient, glances over her shoulder at the ill physician and watches curiously above her mask as the man is racked by another round of coughing. *He looks awful,* she thinks. The hospital doctors agree and immediately move the man into the ICU. In a few days, he will die there.

Saturday, February 22, 2003.
Hong Kong, Special Administrative Region, China

The nephrologist's brother-in-law awakens with a fever. Not too serious, but it frightens him. *It's nothing,* he tells himself. *Nothing.* On the third day he begins to cough, and he takes himself to the hospital. The doctors there go over him thoroughly and assure him it is just a cold. Nothing to worry about.

Forty-eight hours later, the brother-in-law is back in the hospital. This time, to stay. He, too, will die there.

That same day, a nurse who was in the emergency room when the nephrologist was first treated is admitted to the same hospital. A dry unproductive cough, a fever of 104, diarrhea. The three of them, together now, awaiting their fates.

Five days later, the hospital aide who spent the first afternoon with the brother-in-law is also admitted with cough, fever, and diarrhea. And then there were four.

Sunday, February 23, 2003.
Hanoi, Vietnam

An Asian-American businessman steps off the plane that has just arrived from Hong Kong. He is crisply dressed and has had a good trip.

He stayed at his favorite hotel in Hong Kong, the Metropole in Kowloon, and he stayed on the ninth floor. He enjoyed the view from there. Now it is time for family. He feels a little more drained than usual after such a trip, and he feels a little warm even to his own touch, but he will worry about that tomorrow. He collects his luggage and hails a cab.

Three days later, this man is admitted to the private Vietnam French Hospital in Hanoi with cough, fever, and diarrhea.

Week 2. Friday, February 28, 2003.
Hanoi, Vietnam

The Vietnam French Hospital contacts the Hanoi office of the World Health Organization. They have a patient with an unusual influenzalike infection. They ask if someone from WHO could come and take a look. Dr. Carlo Urbani, an infectious-disease specialist, goes immediately to the small private hospital. Perhaps, because it is everyone's worst fear, he suspects a new avian influenza virus. Within a few weeks, he and five other health-care workers will be dead from a previously unknown virus, but it isn't an influenza virus.

Sunday, March 2, 2003.
Hanoi, Vietnam

Doctors put the Chinese-American patient on a ventilator. That helps some, but he continues to deteriorate. On Wednesday, March 5, he is flown to a special hospital in Hong Kong. On the following Wednesday, March 12, he dies.

Sunday, March 2, 2003.
Hong Kong, Special Administrative Region, China

A few blocks away from the hospital that houses the first four Chinese cases of this disease, a Chinese Canadian checks into the emergency room of a second hospital. He has a cough and high fever. For a single day, his stay at the Metropole Hotel in Kowloon, Hong Kong, overlapped with the nephrologist's stay at the same hotel.

Reports of "atypical" pneumonia are moving from hospital to hospital in Hong Kong.

For the next six days, three nurses will attend to the Chinese-Canadian man. On March 3, this man has a severe bout of diarrhea. All three nurses help clean the man and his bedding. None is masked. By March 15, all three nurses develop pneumonialike symptoms and are admitted to this second hospital. As suspicions grow, all three of these patients are transferred to a third hospital, one specifically designated to treat infectious diseases, one especially equipped to handle what looks more and more like a major new threat.

Sunday, March 9, the Chinese-Canadian man's nephew comes to visit him at the hospital. While he is there, one of the nurses treats the ill man with positive-pressure ventilation—forcing air into and out of his lungs. No one thinks to protect himself or herself. Within days, all three nurses and the nephew will also be admitted to this hospital, all complaining of respiratory ailments. Now there were ten in Hong Kong—ten that we knew of.

Week 3. Sunday, March 9, 2003.
Hanoi, Vietnam

The World Health Organization requests an emergency meeting with the Vietnamese vice minister of health. After this meeting, the Vietnamese government quarantines the Vietnam French Hospital, instigates new infection-control procedures in other hospitals, and asks for more expert help from the World Health Organization and the Centers for Disease Control and Prevention in the United States. Doctors Without Borders sends staff members and infection-control suits originally intended for outbreaks of Ebola virus.

Carlo Urbani names the disease severe acute respiratory syndrome — SARS.

Wednesday, March 12, 2003.
Geneva, Switzerland

The World Health Organization issues a global warning about a new infectious disease in Vietnam and Hong Kong. The illness has, at last, caught the world's attention.

Week 4. Saturday, March 15, 2003.
Atlanta, Georgia, United States

The CDC issues a travel advisory to anyone planning travel to Singapore, Hong Kong, or Vietnam.

The World Health Organization issues a heightened global alert and presents the first case definition of SARS. WHO also issues a rare emergency travel advisory to international travelers, health-care professionals, and health agencies.

Week 5. Saturday, March 22, 2003.
Geneva, Switzerland

Singapore reports forty-four cases of SARS to the World Health Organization. At the same time, the Hong Kong Department of Health reveals that two people, who had stayed on the ninth floor of the

Metropole Hotel during that fateful late February and early March, had returned to Singapore.

The Ministry of Public Health in Thailand reports four suspected/ probable SARS patients. Three of these four people had recently traveled to Hong Kong. The fourth was a physician who had treated the Asian-American SARS patient in Hanoi.

The Public Health Agency of Canada reports eleven cases of SARS—one in British Columbia and ten in Ontario. Three of the eleven have already died. The single case of SARS in British Columbia occurred in a man who had stayed at the Metropole Hotel in Hong Kong between February 12 and March 2. Five of the cases in Ontario (Toronto) arose in a single, extended family, and all of those cases appear to be linked to a family member who recently returned from a trip to Hong Kong—a seventy-eight-year-old woman who had stayed at the Metropole Hotel.

Monday, March 24, 2003.
Hanoi, Vietnam

The Vietnamese Ministry of Health reports fifty-nine probable cases of SARS. All of those cases are linked to the hospital where the Asian-American traveler stayed after arriving from Hong Kong.

Tuesday, March 25, 2003.
Hong Kong, Special Administrative Region, China

The Hong Kong Department of Health reports 290 suspected and probable cases of SARS.

The Taiwan Department of Public Health reports six probable cases of SARS. Four of these patients had traveled to the Guangdong Province of China and to Hong Kong during the week before the onset of symptoms.

Wednesday, March 26, 2003.
Atlanta, Georgia, United States

The Centers for Disease Control and Prevention announces that it has received fifty-one reports of suspected cases of SARS in the United States. Four clusters of suspected cases have been identified. All center around a traveler who recently visited Southeast Asia (Guangdong Province, China, Hong Kong, or Vietnam). One of the clusters appears to be attributable to a married couple who stayed at the Metropole Hotel between March 1 and March 6. The man became sick on March 13 after their return from Hong Kong. But his wife didn't become ill until several days later, suggesting that she may have gotten the disease from her husband. Both partners said they remembered that several other guests at the Metropole Hotel were coughing and feverish while the couple stayed there. At the time, they thought little of it.

From China, a web was spreading across the world, a web of disease. At the center of that web, like a fat black spider, sat the Metropole Hotel in Kowloon. From the ninth floor of that hotel, arms now reached to Canada, the United States, Vietnam, Singapore, Ireland, Bangkok, Taiwan, Switzerland, Germany, Italy, Slovenia, Spain, and the United Kingdom. The spider had sprouted those deadly arms in less than a month.

Alarms began to sound. Hoping to save themselves from SARS, people—especially in Asia—covered their faces with surgical masks. Fear reached out from its dark hole and took hold of people's throats from New York City to Tokyo. People, especially people who knew what was really happening, were very afraid.

But then—just as the real picture with all its horrible implications was unfolding its wings across the thin skies of this planet—we got a break, a very big break.

Monday, March 24, 2003.
Atlanta, Georgia, United States

The Centers for Disease Control and Prevention announces it has found a smoking gun. Several laboratories investigating SARS have isolated a previously unknown coronavirus from two patients in Thailand. In the tissues of SARS patients, scientists using the dark beam of an electron microscope found a virus they'd never seen before. The virus was round like a crown—a coronavirus. Shortly afterward, another team of scientists found the same virus in the sera of six more patients. And after that, several more laboratories, working together in a World Health Organization–led investigation, found the same virus in several more SARS patients. The killer became visible.

Coronaviruses are the second most common cause of garden-variety colds. Rhinoviruses (*rhino* for "nose") are the most common causes of colds. But coronaviruses are nothing out of the ordinary, and we've known about them since 1937. They are small, more-or-less round viruses. Under a microscope they appear to be surrounded by a crown.

At one point or another, these viruses infect nearly all of us, but, unlike SARS, most coronaviruses don't sicken us so quickly, and they almost never kill us. Scientists called it the SARS-associated coronavirus (SARS-CoV), and this coronavirus was killing at least 10 percent of those it infected. No coronavirus we knew about had ever done that before.

Week 6. Saturday, March 29, 2003.
Bangkok, Thailand

Carlo Urbani, the physician who first alerted the world to the menace of SARS, dies in a makeshift hospital in Bangkok. For eighteen days he has battled his own infection with SARS. His dedication and efforts to control the spread of SARS saved unknown hundreds of lives but cost him his.

Week 8. Monday, April 14, 2003.
Atlanta, Georgia, United States

The Centers for Disease Control and Prevention announces the completion of the full-length genetic sequence of the SARS-CoV RNA. These data confirm that this is a novel coronavirus. The discoveries are the result of a collaborative effort among scientists at the National Microbiological Laboratory in Canada, the University of California at San Francisco, Erasmus University in Rotterdam, and the Bernhard-Nocht Institute in Hamburg.

Now there was a story to go with the face.

Less than a month after the realization that SARS was a major threat to world health, scientists had isolated the virus and determined its entire genetic sequence—a remarkable accomplishment. Admittedly, we caught a few breaks. SARS-CoV was surprisingly easy to grow in the laboratory. The virus mutated very slowly, so the RNA sequence reported in April was little different from those found in patients with active infections. And scientists cooperated more completely than perhaps they ever had. The first pandemic of the twenty-first century had reared its ugly head and come after us. Surprisingly to many of us, we were ready. Once armed with the RNA sequence, we began to make proteins— pieces of the virus that we might use to immunize and protect the threatened world.

Wednesday, April 16, 2003.
Geneva, Switzerland

The World Health Organization announces that the unusual coronavirus found associated with SARS is the definitive cause of SARS.

Week 12. Thursday, May 15, 2003.
Beijing, China

The Supreme Court of China states that people who violate quarantine for SARS may be imprisoned for up to seven years, and those who knowingly spread the disease may be executed. This statement was issued in response to a physician who had reportedly broken quarantine and fled from Beijing to Linhe in northern China, where he infected one hundred people.

In Beijing, the streets are full of people in white masks, averting their eyes from one another, watching the ground. No one dares to speak to another.

Week 16. Thursday, June 12, 2003.
Toronto, Ontario, Canada

The day of the last probable case of SARS in Canada, and, for the moment, the world. From room 911 of the Hotel Metropole in Hong Kong, SARS had reached across the world, sickened 8,437 people, and killed 813. Then it stopped.

While we fought the clock and the odds, while we mounted unheard of efforts with improbable speed, the pandemic fizzled. In the fading light of that fizzle, there are more questions than answers.

Undoubtedly, humans played a large part in dousing the fires of SARS. Travel restrictions, quarantines, honesty by affected nations, careful and extensive reporting and collaboration—at least in the later stages of the pandemic—all had major effects on slowing the spread of the disease. This was a public health accomplishment unequaled in our history. But something else happened as well, something mysterious. Behind the cloud of that mystery, we dodged a bullet, a microbial bullet that for all the world had seemed to have our names written all over it.

Why, in the end, did Canada have 250 cases of SARS and the United

States only 75? The disease reached both countries about the same time and under similar conditions. Yet in Toronto SARS flourished, while in the United States it foundered.

Where had SARS come from? Many of the men and women who butchered animals for the meat markets of Guangdong Province tested positive for SARS. It seemed the trail led to the meat markets. Where then? Among the animals regularly slaughtered in the meat markets there, Himalayan masked palm civets (a kind of cat), Chinese ferret badgers, and raccoon dogs harbored SARS-like coronaviruses. So early in 2004, the butchers killed thousands of civets. But then other animals—including fruit bats, snakes, and wild pigs—also tested positive for SARS-like viruses. Any, or none, of these could have been the source of the virus that killed 813 people and terrorized another six billion or so.

SARS killed men and women more or less equally, but it favored the old, those over sixty-five. And the virus was fond of health-care workers. Twenty-two percent of those who died in Hong Kong and Guangdong, and more than 40 percent of those who died in Toronto and Singapore, were health-care workers. Many others were equally exposed. What was it about hospitals that helped spread this disease?

SARS leapt from person to person in aerosols exhaled by the sick. But that may not have been SARS' only path to new blood. The virus also appeared in the watery stools of the ill, and it may have moved from there to the mouths of others who were careless with their hands or their drinking water.

And finally, why, in the end, had the SARS epidemic sputtered almost as it was beginning?

In an ironic twist, it appears that it wasn't simply infection with SARS that sickened or killed people. Rather, it was human inflammatory and immune responses that sickened or killed them when these responses moved to destroy SARS-CoV. After infection, it appears that the relatively minor damage to those cells infected by the virus stimulated major inflammatory and immune responses. These resulted in production of large quantities of compounds called cytokines. These cytokines affected blood flow and caused leakage of fluid from the blood vessels. That fluid filled up the lungs of the infected and suffocated them.

Sunday, May 15, 2005.
New York City, New York, United States

The New York Times reports that two and one half years after it began, SARS is officially gone. No cases of SARS have appeared since June of 2003. For the *Times* and a lot of other people, that sounded the death knell for SARS. One more disease that we could bury along those who had died.

But the lack of fresh corpses didn't then and doesn't now mean that the disease and SARS-CoV aren't still out there.

The SARS virus, or another much like it, is bound to reappear. It isn't a question of if. It is simply a matter of when. With SARS we learned what we were capable of—and it's a lot. But we also learned that sometimes a lot isn't enough. In one month, the virus crawled out of room 911 in the Metropole Hotel in Hong Kong and reached around the globe. We responded quickly, but we could never have maintained that level of response. If SARS hadn't stopped itself, only a vaccine would have brought it down. Vaccines now exist. And those vaccines exist only because of the hard work of many people in a very short time and unprecedented efforts and collaboration by public health workers around the world. But it wasn't the vaccines that stopped SARS.

Our technology showed us the reach and the speed of what was happening. In part, our technology saved us, but also in part our technology nearly killed us. The lightninglike flash of SARS around the world relied absolutely on jet airplanes. Without our technology, SARS might have languished in the backwaters of Guangdong Province, might have been known only to the residents of that province, might have frightened or killed only the locals. But that isn't how things happened.

We are a people of many talents and much dedication. But viruses like SARS-CoV have their own brand of determination. We narrowly escaped this one. The next time around, we can only hope to be so lucky.

11 ⠿

Diseases on the Fly: Malaria, Dengue Fever, and West Nile Virus

" 'Over a mosquito,' Sandra said . . . with tears in her eyes. 'That's a hell of a way to lose a loved one.' "

On a hot summer day in August of 2004, Sandra's husband, Wayne Trowbridge, came home from his work in Greeley, Colorado. He was feeling tired and sick. He went to bed immediately. But by the next morning, Wayne's fever was so bad he was trembling.

Sandra told him he had to go see the doctor.

Wayne hadn't been to the doctor in thirty years, and he wasn't of a mind to start now. But four days later he was so sick, he finally agreed to go. He ended up in the ICU of the Northern Colorado Medical Center in Greeley.

Within a few days, Wayne couldn't move his arms or legs. Then his lungs failed. The doctors put Wayne on a ventilator. He lost both his sense of taste and smell, and his weight dropped from 140 pounds to 92 pounds—because of a mosquito bite.

" 'When this hits you so drastic, it just takes your family's whole life,' Sandra said. . . . 'Because all of a sudden, I'm the provider. I'm making all the decisions.' "

The doctors determined that Wayne had a West Nile virus infection. They pulled Wayne through the worst of it, but they kept him in the hospital four months. Finally, he seemed well enough that the doctors moved him to a long-term rehabilitation facility. When asked about his goals, Wayne said all he wanted was to go home and eat steak and mashed potatoes with his wife.

On January 28, 2004, Wayne got his wish. The doctors sent him home and Sandra fed Wayne steak, mashed potatoes, and gravy. Wayne loved it.

But just a few days later, on February 8, Wayne's lungs began to fill with liquid. A helicopter took him back to the hospital.

"That morning, I knew he would never come home again," Sandra said.

West Nile Virus and Encephalitis

A virus killed Wayne Trowbridge, West Nile virus. A mosquito gave the virus to Wayne and a crow or some other bird probably gave the virus to the mosquito. None of this, of course, makes any difference to Wayne. But it should to the rest of us.

West Nile virus was first isolated from a feverish woman in the West Nile district of Uganda in 1937. The virus flared again in Egypt in the early 1950s and in Israel in 1957. In the 1960s, West Nile first showed up in horses in Egypt and France. Sixty-two years after its discovery in Uganda, the virus turned up in New York City.

First it took the birds. In New York City, it rained crows. Then the virus went looking for horses and humans. In August of 1999, five patients with confusion, fever, and weakness showed up in the intensive care unit of the Flushing Hospital of New York. Four of the five lost the ability to breathe and had to be placed on ventilators. Shortly afterward, four more patients with less severe symptoms were diagnosed.

By the year 2000, West Nile virus had sickened eighty-three people and killed nine of them.

Never before seen in the western hemisphere, West Nile virus was about to become a household commodity.

We don't know for sure how the virus came to the United States, but the strain of virus that caused most cases in the United States looks a lot like West Nile virus from Israel. Regardless of where it came from, the virus did not arrive inside of a human being.

Humans get the disease when a female mosquito (only the females bite), laden with West Nile virus, bites one of us. As the mosquito feeds, it injects saliva to keep its unwilling host's blood from clotting. To aid in its journey from animal to animal, West Nile virus has learned how to get inside of mosquito salivary glands. So when an infected mosquito bites us, along with a little bit of mosquito spit comes a whole lot of West Nile virus—directly into our blood.

The mosquito (usually of the *Culex* variety in North America) that works that magic gets the virus from birds, most often corvids—crows, ravens, magpies, jays. These birds are the best sources of virus for mosquitoes, because after infection, the blood of these birds teems with virus. In 2001, just two years after the virus first showed up in New York, 359 counties in 27 states in the United States reported West Nile virus activity. Ninety-one percent (328) of those counties reported a total of more than 7,000 dead West Nile virus–infected birds, 5,154 of them crows. Overall, more than half of the nearly 10,000 crows tested were positive for West Nile virus. West Nile virus was decimating the crow populations.

"I've seen crows looking great, eating and behaving normally, for five or so days after infection, and then literally drop dead within hours," says Dr. Richard Bowen, a veterinarian and a West Nile virus researcher at Colorado State University. "The levels of virus in the blood of these birds are somewhere between one hundred thousand and one million times the levels found in horses." That amplification makes the birds a reservoir of virus for the mosquitoes. "When we look for virus in these birds, it's everywhere, even in heart muscle and in the birds' eyes. In the end, I think these animals may actually die from heart failure, their viral load is so great." Dr. Bowen has looked for West Nile virus in a lot of species, and he has found it in most of them, but nothing like the levels of virus seen in crows and other birds.

"The birds grow virus like nothing else."

Fortunately, that sort of viral amplification doesn't happen in humans and other mammals. The virus doesn't multiply as easily in our blood, nor does it infect so many of our tissues. Because of that, after West Nile virus infection, we don't die from heart failure, and we rarely transmit the virus directly to one another—at least not via mosquitoes. We need birds for that. So West Nile virus came to North America inside of a mosquito or a bird.

West Nile virus is a flavivirus. Flaviviruses are a large family of insectborne viruses that include the viruses that cause yellow fever, dengue hemorrhagic fever, St. Louis encephalitis, and West Nile encephalitis. All of these viruses can affect humans' nervous systems and can, under the right conditions, cause encephalitis—an inflammation and swelling of the brain.

Once a mosquito deposits West Nile virus in human blood, the virus heads for human brains.

About 80 percent of people infected with West Nile show no overt signs of infection. Somehow, before the virus can do any real damage, these people's immune systems corral the virus and eventually destroy it. A small proportion of people develop flulike symptoms—fever, headache, body aches, nausea, vomiting, and sometimes swollen lymph nodes or a skin rash on the chest, stomach, or back.

In a few of these people, the virus crosses the blood-brain barrier and infects some of the cells of the nervous system in the brain or spinal cord or both. The infected cells trigger inflammatory and immune responses. As our defenses move to destroy the virus, the blood vessels in our brains expand. Fluid begins to leak out of our blood vessels, our white blood cells release chemicals that further stimulate inflammation, and our brains or our spinal cords swell.

Normally, swelling isn't a big problem. A finger or an ankle swells and with a little ice, things get better. But bone surrounds all of the central nervous system (brain and spinal cord). When the brain or spinal cord swells, there is no way to release the excess pressure. Swelling in the central nervous system causes pressure inside the spine or skull. That swelling can reduce blood flow to nerve cells. Nearly 30 percent of the oxygen we inhale goes to our brains and our spinal cords. So when the blood flow to the central nervous system decreases, bad things begin to happen, quickly. Nerve cells die and systems start to fail—legs, arms, lungs stop doing the things they used to do.

About 1 percent of those infected with West Nile virus eventually develop meningitis (an inflammation of the meninges, the membranes that surround the brain and spinal cord) and/or encephalitis (an inflammation of the brain itself). That inflammation also puts pressure on the brain and/or spinal cord and causes loss of nerve function. The symptoms of encephalitis and meningitis include headache, high fever, stiff neck, muscle weakness, stupor, disorientation, coma, tremors, convulsions, paralysis, and sometimes death. Often, treatment requires extended hospitalizations and long periods on ventilators that assist with breathing. Rehabilitation is a long, lonely, and painful process.

Patricia Heller, a young Colorado woman, skied whenever possible and bicycled competitively. But one afternoon, after one of her long bicycle trips, she collapsed in the street next to her bike. She wrote it off as cramps or dehydration. But the next day, she could not move her left leg. The doctors diagnosed it as a West Nile virus infection. Obviously from a mosquito bite, though Patricia recalled none. The young woman underwent months of agonizing therapy, trying to regain what the virus had taken from her. But two years later, she still had not fully recovered. The damage done to the cells of her spinal cord or brain did not just go away. Nerves take years to heal, if they heal at all.

For Patricia and many others, West Nile virus encephalitis is no "mild" disease, as public health officials have sometimes described it. Neither is it a comfort to the infected people that the odds were in their favor when the mosquitoes bit them.

With proper medical treatment, most of these people, like Patricia, survive. A few, a significant few, like Wayne, do not. For these few, infection with West Nile virus becomes a mosquitoborne nightmare and a death sentence.

By 2001, just two years after the first cases of West Nile appeared in New York, horses in New Hampshire, Massachusetts, Connecticut, New York, New Jersey, Delaware, Maryland, Virginia, North Carolina, Florida, Louisiana, Mississippi, Georgia, Arkansas, Kentucky, Tennessee, Indiana, and Illinois tested positive for West Nile virus. The same year, Massachusetts, Rhode Island, Connecticut, New York, New Jersey, Pennsylvania, Maryland, Georgia, Alabama, and Florida all reported human infections by West Nile virus. Eighteen people were dead.

The country was catching fire.

By 2002, Texas, Oklahoma, Colorado, Nebraska, Kansas, Missouri, Iowa, Minnesota, Wisconsin, Michigan, North Dakota, South Dakota, Montana—the entire midsection of the country—and California—also reported cases of West Nile virus infection. The same year, 4,156 human cases of West Nile virus infection were reported and 284 people died.

In maps that show the animal cases of West Nile virus infection in 2002, the country now bled red from the East Coast to as far west as Wyoming. It took the virus only three years to open that wound.

By 2003, every state in the contiguous forty-eight, except Washington and Oregon, reported cases of West Nile illness in humans. That year, West Nile virus infected at least 9,862 people and killed 264 of them. But the number of cases was beginning to fall along the eastern seaboard. The wave had crested as it rolled west.

Why the number of cases in the eastern United States began to drop isn't clear. Human immunity may play some role, since many people survived West Nile virus infection and became immune to further challenge. But humans are not the reservoir for West Nile virus; birds are. Only birds have high enough levels of virus in their blood to efficiently transfer the infection to mosquitoes. So human immunity alone would not be enough to bring West Nile infections to such a screeching halt. Some birds must also have developed immunity to West Nile virus. That would dramatically drop the number of birds capable of transferring West Nile virus to mosquitoes. That, in turn, would slow the transmission to horses and people and cats and dogs and every other susceptible creature.

Regardless of the cause, the wave rolled west. And Colorado got hit the hardest.

In 2003, Colorado reported 2,947 cases of West Nile infections in humans—nearly as many as the rest of the country combined. Sixty-three of those people died. No other state had it so bad. Nothing about Colorado or the weather or anything else in 2003 obviously accounts for this plague. As bad as those figures sound, though, research indicates that things were actually much, much worse. Blood tests done in 2004 and 2005 show that West Nile virus infected many more people than we had suspected, maybe as many as 70 percent of the people in Colorado during 2003 and 2004 and most of the dogs, cats, horses, sheep, and every other mammal tested, as well as most corvids, other birds, and many reptiles. As West Nile virus rolled across this country, few were spared.

By 2004, the virus had missed only the state of Washington among the contiguous states in the United States. Reports of the disease among eastern and central states had fallen off dramatically.

As of September of 2005, thirty-two states reported 821 cases of human West Nile virus illness. Eighteen cases were fatal. For comparison, in 2004, a total of 1,191 West Nile virus cases had been reported as of September 7, 2004.

About 2,300 corvids and 500 other birds had died from West Nile virus infections by September 2005. Twenty-eight states also reported West Nile infections in horses, as well as infections in three dogs in Minnesota and Nebraska, four infected squirrels in Arizona, and infections in two unidentified animal species in Arizona and Illinois. West

Nile infections appeared in 549 sentinel chicken flocks from eleven states and in one sentinel horse in Minnesota. (Sentinel flocks are flocks of animals maintained by local public health services to monitor the progress of infections.) And in 2005, Alabama, Arizona, Arkansas, California, Colorado, Connecticut, Florida, Georgia, Idaho, Illinois, Indiana, Iowa, Kansas, Louisiana, Maryland, Massachusetts, Michigan, Minnesota, Missouri, Montana, Nebraska, Nevada, New Jersey, New Mexico, New York, Ohio, Oklahoma, Oregon, Pennsylvania, South Carolina, South Dakota, Tennessee, Texas, Utah, Virginia, and Wisconsin combined reported a total of 6,833 West Nile virus–positive mosquito pools. As of late 2005, the virus was still clearly thriving. And the virus flared once again in the summer of 2005. But all the totals, except for the number of infected pools of mosquitoes, were down significantly from 2004.

The flood appeared to have crested. Once again, it looks like we dodged a bullet, or at least some of us did. If this virus had caused frequent, severe illnesses, we and most of the animals that surround us would not be here today. The country, from New York to California, would be littered with our dead.

Imagine a disease that swept through this country like West Nile virus but, unlike West Nile, it sickened or killed even 1 percent of those infected. That would have been two million corpses. We would have been struggling to find the people and the space to bury all of our dead.

There is no doubt that education about repellants and spraying for mosquitoes made some difference in the spread of West Nile virus, maybe even a lot of difference. But our biggest break came from the virus itself. Even though the virus infected many of us, most of the infected never even noticed.

It could have been horrendous. But since it wasn't, most people are willing to move West Nile onto the far-back burner of human concerns.

That might be a mistake.

"Still, there are more cases of paralysis [from West Nile] than there were in many years of polio. [And] last year about one third of the West Nile cases reported to the CDC had neurological complications, like meningitis or encephalitis," says Dr. Henry Masur of the National Institutes of Health. The fire still smolders; for some it flares.

Wayne Trowbridge was twenty-three and Sandra just eighteen when the couple eloped. They were madly in love, and soon they had six daughters. Then, one day, the family picked up and moved to Colorado. It seemed like it was the best place for them. Thirty-five years after they married, Wayne died from an infection with West Nile virus—killed by a mosquito bite in Greeley, Colorado.

Sandra told reporters that "Wayne was a simple man, and he wanted simple things from life. He wanted a comfortable fun life, and he wanted everybody to have fun with him."

Wayne thought of most things it would take to make all of that happen for him and Sandra, but he overlooked mosquitoes. Most of us do.

For most of us, West Nile virus is fading into an unremarkable and unimportant past. Not so for Sandra Trowbridge. She said, according to one newspaper: "West Nile is not over with. There are too many people laying in hospitals and rehab centers. People better start thinking about next summer."

Malaria

People in the United States haven't always been so careless about mosquitoes. But we humans are quick to forget—sometimes, maybe too quick. West Nile encephalitis is not the first disease to be unleashed on large parts of the United States through the lances of mosquitoes. Malaria holds that honor.

"From colonial times until the 1940s, malaria was *the* American disease. At the dawn of the twentieth century, it thrived from New York to Florida, from North Carolina to California. Up to 7 million Americans were stricken by it every year until the mid-1920s, and 3,900 people died from malaria in the United States as recently as 1936," writes

Robert Desowitz in *The Malaria Capers.* One of the Continental Congress' first expenditures in 1775 was for $300 worth of quinine to protect George Washington and his troops from malaria. Eighty or so years later, during the Civil War, nearly half of the white soldiers and three-quarters of the black soldiers contracted malaria annually. For hundreds of years, North America was wretched with malaria.

At least three species of mosquitoes capable of transmitting malaria have existed for centuries in the United States. But since there is little evidence of malaria in the bones of Native American peoples buried before the arrival of Columbus, it appears that malaria came to this country along with colonists or their slaves. Once here, the disease thrived. Nearly everything about North America suited malaria just fine, and it killed thousands of us.

By the middle of the twentieth century, though, malaria was pretty much gone from the United States. Its disappearance was mostly due to DDT and mosquito-eradication programs—though window screens and the movements of people from the swampy bottomlands of the South into urban centers also played a part. Maps showing the distribution of malaria in the United States toward the end of the nineteenth century paint the entire midsection, the Atlantic seaboard, California's central valley, and parts of Washington and Oregon black with the disease. But by 1932, malaria had dwindled to a few southern states, and by 1943, the country was once again virtually free of malaria. Outbreaks of malaria continue to occur in this country, some of them locally acquired (meaning they didn't come here from outside the United States). But so far, each of these outbreaks has been quickly controlled using drugs and mosquito-control policies.

Because of that, only a few decades after millions of illnesses and thousands of deaths, North Americans have forgotten about malaria. Malaria is a disease of the tropics. Now we look at mosquitoes as annoyances, not killers.

For much of the rest of the world, the story is very different. In the underdeveloped parts of this world, particularly sub-Saharan African, malaria is a leading cause of death. Every year, malaria acutely infects more than four hundred million people and kills more than a million— mostly children or pregnant women, and mostly in Africa, but also in India, Asia, and South and Central America. Many more people are chronically infected with malaria. Worldwide, approximately 2.5 billion people are at risk for the disease. Malaria has reshaped the face of the globe. Not only are countries with great poverty at greater risk for malaria, but countries with malaria are at greater risk for great poverty because of the tremendous drain this disease places on countries' economic reserves.

Parasites cause malaria. Parasites are single-celled or multicellular animals that, of necessity, live inside the bodies of other animals. Parasites make their livings by co-opting the machinery of others' healthy bodies for their own parasitic purposes. By definition, through this process the parasite gains and the host loses, because, in the process of making its living, the parasite kills a lot of its host's cells.

Parasites cause horrible diseases like malaria, elephantiasis, yellow fever, and dracunculiasis (Guinea worm disease). People get Guinea worm disease from drinking infested water and swallowing the fleas that carry the parasitic worm. A year or so later, adult worms begin to emerge through these people's skin, worms as big around as strands of spaghetti and two to three feet long. The only treatment is to slowly wind the worms onto sticks for the next few weeks until they can be pulled out and killed. If you don't get all of the worms, they'll be back in a year or so.

The parasites that cause malaria all belong to the *Plasmodium* family—*Plasmodium falciparum, Plasmodium vivax, Plasmodium ovale, Plasmodium malariae. P. vivax* and *P. falciparum* are the most common and *P. falciparum* is the most deadly.

Mosquitoes move the parasites from infected people to uninfected people, in particular mosquitoes of the genus *Anopheles*. Unlike West Nile virus, the malaria parasite reaches high enough levels in the blood of infected people so that mosquitoes can transfer malaria from person to person. Like West Nile virus, *P. falciparum* and the others have learned to invade mosquitoes' salivary glands. So when an infected mosquito bites, along with saliva, the mosquito injects the malaria parasite directly into the blood.

Once in the blood, the parasite travels to the liver, where it infects some of its cells. Inside those cells, the *Plasmodium* multiplies until some of the liver cells rupture and release the parasite into the blood. There, the malaria parasite infects red blood cells. Inside red blood cells, the *Plasmodium* replicates again until many of the blood cells rupture and release large numbers of new parasites that infect more red blood cells. At this point, if a mosquito feeds on an infected person, the mosquito will become infected and can then transmit the disease to other humans.

As the parasite reproduces in the red blood cells, it causes several of the symptoms of malaria—fever spikes alternating with chills, headaches, aches and pains, diarrhea, and abdominal pains. Different parasites cause different patterns of fever. But generally, the fever correlates with rounds of replication and red-cell rupture that the parasites cause.

In children, the effects of malaria are especially severe. In infected children, the parasite can also cause cerebral malaria, an often fatal complication of the disease. Cerebral malaria occurs when the parasites affect people's brains. And cerebral malaria is especially common in areas where malaria infects the majority of the mosquitoes and children get bitten hundreds of times each year. This type of malaria seems to result from parasite-induced changes in red blood cells and in the blood vessels in the brains of these children. This results in reduced delivery of oxygen to the children's brains and death of nerve cells, which can quickly lead to the death of a child.

As mentioned earlier, malaria infects somewhere between three hundred million and five hundred million people living in sub-Saharan Africa. That is five times more people than are infected with HIV, measles, tuberculosis, and leprosy combined. One out of every four childhood deaths in southern Africa is due to malaria.

Malaria can be treated with some drugs that slow the disease and may cure it. But malaria parasites in many regions of Africa are becoming resistant to what have been the safest and most effective treatments.

In the future it seems unavoidable that malaria will spread and take an even greater toll on human beings.

Because of the continued presence of *Anopheles* mosquitoes in the United States, there are recurring outbreaks of malaria in this country. And though control efforts so far have been remarkably effective in limiting the spread of these outbreaks, there is still the potential for significant impact on North American people.

Beyond its terrible toll on life in Africa, malaria also causes more than $12 billion in lost gross domestic product on that continent, even though the disease could be controlled for less than a tenth of that amount. The shape of this world—who is on top and who is on the bottom; who will live and who will die—is in part due to malaria and *Anopheles* mosquitoes.

Ironically, we could change most of this if we cared to. We almost completely eliminated malaria from the United States using insecticides, DDT in particular. There is little question that DDT could still have a great impact on malaria in Africa. Ecuador has reduced malaria infections by 60 percent using DDT. Ecuador's neighbor, Bolivia, however, bowed to international pressure and banned DDT. During the same period of time when malaria infections were decreasing in Ecuador, Bolivians saw an 80 percent increase in malaria infections. DDT is cheap, and even where some mosquitoes have developed resistance, DDT appears to be an effective repellant.

Rachel Carson's *Silent Spring* was an eloquent condemnation of DDT. And many of our predatory bird populations have still not recovered from the effects of widespread agricultural spraying of DDT. But that is not what might happen from the limited use of DDT in Africa.

In Africa, and other poor areas, many people live in substandard housing, sometimes even mud huts, all of which lack window screens or curtains. In these dwellings, spraying the interior walls with DDT has been shown to help reduce the number of mosquito bites and infections with malaria.

This sort of focused spraying of DDT has never been shown to have any dramatic effects on either people or the environment. Most of the effects of DDT we are familiar with appear to have been due to large-scale agricultural use of the insecticide.

" 'It is inconceivable, unconscionable and reprehensible that the European Union [EU] would put its fear of pesticides above the lives of innocent Ugandan mothers and children,' the Congress of Racial Equality's Cyril Boynes Junior said today. 'But that is exactly what is happening. EU chargé d'affaires Guy Rijcken's vile threat is an abuse of his authority and a serious human rights violation. The Government of Uganda should immediately review his diplomatic credentials,' " reported the *Canada Free Press*.

Boynes was responding to Rijcken's memo that asked the Ugandan government to ban the use of DDT or risk losing its current exports to the EU. Rijcken's reasons had been that DDT posed serious health and environmental risks. Boynes is the international director of the Congress of Racial Equality in the United States.

Boynes's wife, Fiona, is Ugandan, and she herself has contracted malaria more than once. " 'Fiona Kobusingye lost two sisters, two nephews, and her own son to this disease,' [Boynes] noted. 'Don't talk to me about birds,' she has said. 'And don't tell me a little DDT in our bodies is worse than losing more children to malaria. African mothers would be overjoyed if our biggest worry was DDT in our breast milk.' "

Malaria comes from the Italian *mal aria*, or "bad air." People once believed that the disease came from breathing the fetid air over the swamps in Europe and Africa. It was the mosquitoes, of course, that came from the swamps, not the foul air, which brought malaria. But the name remains, as does *Plasmodium* and *Anopheles*.

Between 1957 (when modern surveillance methods began) and 1994, seventy-seven cases of "cryptic" malaria were reported from twenty-one of the United States. Cryptic malaria refers to cases of malaria that were not obviously acquired while outside of the United States. Of these seventy-seven cryptic cases, seventy-four most likely resulted from bites by mosquitoes infected in the United States. The mosquitoes still thrive here. All they need is an infected body to bite.

International air travel will surely continue to provide the bodies. The mosquitoes will do the rest.

And of course, in much of the rest of the world, airplanes, new mosquitoes, and new bodies are unnecessary. Everything essential to malaria is present every moment. Every day in sub-Saharan Africa, three thousand children die from malaria—a boy or a girl every thirty seconds, every day, every week, every month, every year.

Dengue Virus and Breakbone Fever

About the same time that George Washington was feeding his men quinine to prevent malaria in the American colonies, there was an outbreak of a new disease in temperate and tropical areas in Asia, Africa, and North America—including Philadelphia, the home of the Continental Congress. The people coming down with this disease had high fevers, their eyes hurt, they were nauseated and vomiting, and their joints hurt so much it felt as if their bones were breaking. "Breakbone fever," they called it. Over time, most of those infected got better. But later, some of these people developed more serious diseases, especially the children. For no apparent reason, they started to bleed from their gums and their noses, and blood pooled under their skin until they had huge purplish bruises. Sometimes, the diseases got even worse, and infected people began to bleed massively, collapsed, and lost consciousness. The men of medicine rubbed their temples, and the shamans prayed. The diseases disappeared for a while, but always they came back. The Swahili had already named the terrible fever "Ka-Dinga pepo"—a sudden cramp, an evil spirit. Later, Ka-Dinga pepo became "dengue."

In the 1800s, people discovered that mosquitoes transmitted dengue. But the virus was not identified until the 1940s, when the Japanese and Americans were fighting in the Pacific and Asia, and large numbers of their men were falling to dengue. By 1956, scientists knew there were actually four slightly different forms (called serotypes) of dengue virus— dengue-1 (DEN-1), dengue-2 (DEN-2), dengue-3 (DEN-3), and dengue-4 (DEN-4). It was also clear that there were three different forms of the disease. The most common form was classic (or uncomplicated) dengue characterized by high fever, severe headache, pain behind the eyes, severe joint and muscle pain, nausea and vomiting, and a rash. In addition, there was dengue hemorrhagic fever (DHF), with symptoms that included serious damage to lymphatic and blood vessels and bleeding from the nose and gums and under the skin. And worst of all, there was dengue shock syndrome (DSS), the result of massive bleeding and fluid leaking out of the blood vessels, which led to fatal shock.

It still isn't entirely clear what is responsible for the different forms of the disease, but the incidence of dengue hemorrhagic fever is increasing dramatically. In 1981, an outbreak of dengue fever in Cuba affected more than 344,000 people. More than 10,000 of these people developed dengue hemorrhagic fever and 158 of them died, including 101 children. Another outbreak followed in 1997. This time the virus infected nearly 3,000 people and included 205 cases of dengue hemorrhagic fever and 12 deaths. Dengue fever returned once more, to Cuba in 2001–2002. This time dengue infected at least 14,443 people, with 81 cases of dengue hemorrhagic fever and 3 deaths.

Currently, about five hundred thousand cases of dengue hemorrhagic fever require hospitalization every year. In 2002, Rio de Janeiro alone reported nearly three hundred thousand cases of dengue fever, more than fifteen hundred of which were dengue hemorrhagic fever. Most of those affected were children. And this outbreak spread to Cuba. Even with proper medical care, 2.5 to 5 percent of these people die. Without medical treatment, fatalities from dengue hemorrhagic fever may be as high as 20 percent.

People who survive an infection with one dengue virus serotype are immune only to that one serotype. They may still be infected by any of the other three. So it is possible to have dengue fever more than once. There is some evidence that people who survive a first dengue virus infection are more likely to have dengue hemorrhagic fever following a second infection. Since more people are becoming infected with dengue and surviving, perhaps it is understandable that the number of cases of dengue hemorrhagic fever is rising.

From the 1950s on, in parts of Southeast Asia including Thailand, Indonesia, and Malaysia, dengue hemorrhagic fever was a continuing public health issue. But in nearby Sri Lanka and India, even though all four serotypes of dengue were present, dengue hemorrhagic fever was extremely rare. All of that changed in 1989, when dengue hemorrhagic fever suddenly spread through Sri Lanka and India. Nothing about the mosquitoes that carry dengue had changed. It had to be the virus itself that was different. Early studies quickly confirmed that. A mutant form of DEN-3 infected most or all of the hemorrhagic fever patients in Sri Lanka and India. A change in the virus changed the disease, suddenly and significantly.

Interestingly, the same studies that revealed the DEN-3 mutant in Sri Lanka and India showed that the identical mutant was responsible for outbreaks of dengue hemorrhagic fever from Mexico to Brazil in the Americas and in Mozambique. A DEN-3 mutant that most probably originated somewhere in the Indian subcontinent had found its way to East Africa and Sri Lanka in the 1980s and from there to Central America in the 1990s. Since then, there have been sporadic outbreaks of dengue hemorrhagic fever in southern Central America near the Mexican-Guatemalan border.

In part, the outcome of infection—an annoyance or a dance with death—has to do with the virus. But the outcome of infection also has to do with the person. Not only are people who have been previously infected more likely to develop dengue hemorrhagic fever, but genetics also makes a difference. Some people, simply because of who they are, are more likely to develop severe disease after even a single dengue virus infection. Like so many others, dengue is a complex disease.

Our ways of living make it even more complex. After a relatively few years, the genetic accident that changed everything in the backwaters of India showed up on the doorstep of Texas. Jet airplanes are changing more than just business schedules.

Today, the World Health Organization estimates that there are 50 million to 100 million cases of dengue fever per year. In 2001, nearly a million of these cases occurred in the Americas. More than 2.5 billion people—nearly one half of the world's population—are now at risk for dengue virus infection. That is because all of these people live within the reach of two types of mosquitoes—*Aedes aegypti* (far and away the most common vessel for dengue virus) and *Aedes albopictus*, the Asian tiger mosquito.

Once, many fewer people lived within the natural habitats of these mosquitoes. In the 1950s and 1960s, the Pan American Health Organization, which included the United States, Central America, and South America, undertook a major effort to eradicate *Aedes aegypti* using DDT and other insecticides. At the time, the aim of this program was not dengue virus but yellow fever virus, also carried by *Aedes aegypti*. This effort was remarkably successful in all countries except the United States where, in spite of all efforts, dengue continued to plague the Gulf Coast states, including all of Florida. By 1970, Central and South America, though, were free of *Aedes aegypti*. Another remarkable public health accomplishment.

By 1997, most of the programs designed to limit the range of *Aedes aegypti* had ceased, primarily because of lack of funds—due either to redirection of these funds into higher-priority projects or increasing poverty. Regardless of the reason, by the end of the twentieth century, essentially no one was any longer seriously involved in mosquito control in the Americas. And that is still the case.

According to the Centers for Disease Control and Prevention:

Considerable emphasis for the past twenty years has been placed on ultra low volume insecticide sprays for adult mosquito control, a relatively ineffective approach for the control of *Aedes aegypti*. Second, major global demographic changes have occurred, the most important of which has been uncontrolled urbanization and concurrent population growth. These demographic changes have resulted in substandard housing and inadequate water, sewage, and waste management systems. . . . Third, increased travel by airplane provides the ideal mechanism for transporting dengue viruses between population centers of the tropics. . . . Lastly, in most countries, the public health infrastructure has deteriorated. Limited financial and human resources and competing priorities have resulted in a "crisis" mentality with emphasis on implementing so-called emergency control methods in response to epidemics rather than on developing programs to prevent epidemic transmission. This approach has been particularly detrimental to dengue control because, in most countries, surveillance is [just as it is in the U.S.] very inadequate.

Aedes aegypti has not only returned to all of the areas once cleared; it has now spread to many new areas. As anyone could have predicted, dengue virus has begun to reemerge in many parts of the Americas. Though dengue virus is a formidable foe, it has reemerged as a major threat to world health mostly because of what humans have done.

Oddly, so far, yellow fever has not reemerged as a major disease, but it seems likely that it will.

Dengue is now endemic in more than one hundred countries in Africa, the Americas, the eastern Mediterranean, Southeast Asia, and the western Pacific. Southeast Asia and the western Pacific are currently the most seriously affected. And each year, dengue infects another one hundred million people. Today, dengue virus is the most important mosquitoborne viral disease in the world. Nothing else really comes close.

But we are still far from understanding dengue viruses and the diseases they cause. Current treatments for dengue virus infections are purely palliative. There is no cure, and there are no vaccines.

Before 1970, only nine countries reported epidemics of dengue hemorrhagic fever. By 1995, that number had quadrupled. By 2001, there were more than six hundred thousand cases of dengue infection in the Americas, and fifteen thousand of these were dengue hemorrhagic fever—a twentyfold increase since 1965. In 2005, dengue fever reemerged in Hawaii after an absence of more than sixty years, sickening somewhere between 122 and 1,650 people. Still, surveillance efforts for dengue virus in the United States and other countries remain minimal to nonexistent.

Again, according to the Centers for Disease Control and Prevention:

Pakistan first reported an epidemic of dengue fever in 1994. The recent epidemics in Sri Lanka and India were associated with multiple dengue virus serotypes, but DEN-3 was predominant and was genetically distinct from DEN-3 viruses previously isolated from infected persons in those countries. After an absence of 35 years, epidemic dengue fever occurred in both Taiwan and the People's Republic of China in the 1980s. [. . .] Singapore also had a resurgence of dengue/DHF [dengue hemorrhagic fever] from 1990 to 1994. [. . .] In other countries of Asia where DHF is endemic, the epidemics have become progressively larger in the last 15 years.

[But] the emergence of dengue/DHF as a major public health problem has been most dramatic in the American region.

I don't suppose anyone ever expected to completely eradicate dengue and West Nile viruses or Plasmodium or the mosquitoes that carry it. But after the great strides toward mosquito control in the 1950s, 1960s, and 1970s, people were hopeful. No one could have imagined how dramatically these diseases would expand at the end of the twentieth and beginning of the twenty-first centuries. Medicine had, after all, worked wonders against so many threats. Nor did any of us in the developed world ever imagine we would come to fear the sting of the mosquito's barb so soon or so deeply. But we might have expected that someone would continue to try to limit the spread of disease-ridden mosquitoes and to monitor the United States, at least, for the appearance of disease that could sicken millions of us. We might have expected that.

12 ⦂⦂

Agents of Change: Anthrax, Plague, and Bioterrorism

On the morning of September 19, 2001, Robert Stevens rolled into the parking lot in his shiny four-wheel-drive pickup truck, fishing rods rattling in the gun rack behind his head. Robert grabbed his coffee from the cup holder, jerked open the door, and climbed out. As he did, he forgot his jacket with his glasses in the breast pocket. He did that many mornings. This morning, though, it would cost him his life.

Robert was sixty-three years old. He was a photographic editor for *Sun* newspaper, a supermarket tabloid in Miami, and thinking of retirement, but he was still robust for sixty-three. Robert was responsible for creating pictures that made it appear that people are doing things that they are not—aliens with Elvis, that sort of thing.

On his way to his office, he grabbed a handful of envelopes from his box in the mail room.

In his office, Robert sat and began sorting through his mail. Mostly it was just the usual stuff that morning, but one envelope stood out. He ran his hand through his thinning hair and took a long look at the crudely addressed envelope. He held it close to his face so he could read the writing. Block letters, no return address. With his letter opener, Robert slit the envelope and pulled out the letter. A fine white powder wafted off the paper and into the air around Robert's head. Surprised, Robert recoiled and inhaled.

Some prank, he thought and tossed the whole thing into the trash. He reached for the next envelope. But even as he reached out his arm, Robert was a dead man.

Less than two weeks later, Maureen Stevens brought her husband to the hospital emergency room. Robert was delirious and completely disoriented. The staff quickly admitted him to the hospital and then to ICU. That night he lapsed into a coma.

The doctors tried every test they could think of on Robert and his blood but found nothing. The last thing they tried was a spinal tap. That showed something. So the doctors' preliminary diagnosis was meningitis.

"It's bad, but I thought it was something you could get over," said Mrs. Stevens.

Two days later, Robert died. The final diagnosis was inhalation anthrax. Anthrax—black bane, the fifth plague, woolsorter's disease—a bacterium that for centuries only cattlemen and sheep ranchers worried about, had found its way into the mail and from there into a human being's lungs.

Once the FBI launched its investigation, it became clear that Robert Stevens was not the first person to get anthrax from the envelopes of terrorists. That honor probably belongs to Johanna Huden, a thirty-one-year-old editorial assistant at the *New York Post* newspaper. The letter was addressed to the *Post*'s editor in chief, Col. Allan. Huden often opened the editor's mail.

Later, Huden noticed a blister on her right middle finger. She figured it might be a bite or something similar. But the next day, while attending a wedding, she idly scratched at the blister and it ruptured, spreading a creamy liquid across her hand.

Her doctors studied her on and off for weeks but could not figure out what was wrong with Johanna's finger. Plastic surgeons chopped off pieces of the finger, black, dry, dead. But the ugly black eschar always came back. Finally, it was news articles about other cases of anthrax and friends that gave Johanna her first clue about what was eating her finger—cutaneous anthrax. An eschar, the dead and blackened skin at the site of infection, is a classic sign of cutaneous anthrax. A few weeks and a lot of the antibiotic ciprofloxacin later, Johanna recovered.

Robert Stevens wasn't so lucky. Though he wasn't the first infected by a bioterrorist's anthrax, he was the first to die because of it.

The last to die, in 2001, was Ottie Lundgren, a ninety-four-year-old woman living alone in southwest Connecticut. Ottie never did any-thing, by any stretch of the imagination, to deserve what she got. Never-theless, at 10:35 on the morning of November 21, she died of inhalation anthrax. Her only crime had been that her mail had crossed paths with an anthrax-laced letter intended for Senator Tom Daschle or Senator Patrick Leahy.

Ottie was just collateral damage.

As of November 21, 2001, there were twenty-three known cases of anthrax infection, and five people had died. Bioterrorism, something everyone had feared but most hoped would never happen, now leered at us from Florida and Washington, DC, from New York City and New Jersey. Only weeks after the savage attacks on the World Trade Center, the war had moved from using planes as big as buildings for bombs to using the microscopic as murderers. Our world shook.

Terror fell from the skies and it blew in from the streets. Most of us had an abiding sense that this was only the beginning. The person (or persons) responsible for killing Robert and the others, as well as sicken-ing at least sixteen more, is (are) still at large and unknown. And people still fear their mail and the dusting of white powder that littered Robert's desk.

The lines of battle had shifted from the back streets of the Middle East and Afghanistan to the main streets of the United States, and the weapons of that war had spread their wretched arms to include the bio-logical.

Anthrax

Anthrax was probably the fifth and sixth plagues spoken of in Exodus in the Old Testament, some fifteen hundred years before the birth of Christ. The fifth plague was anthrax in animals, killing the livestock. The sixth plague—boils—was cutaneous anthrax in people infected by the animals.

The plague Homer spoke of in the *Iliad*—Apollo's burning wind of plague—might well have been anthrax. And Virgil's *Georgics*, written in about 29 B.C., provide a detailed description of anthrax in the section on farming.

As mentioned earlier, anthrax was the disease that Robert Koch used in the nineteenth century to develop his famous postulates for the proof that a biological agent causes a particular disease. And it was Pasteur's vaccine against anthrax in sheep that first convinced people of the reality of microscopic life.

Anthrax has been with us for a long time. But for nearly all of that time, anthrax was not a particular problem in humans, although humans did, on occasion, contract the disease. In humans, anthrax comes in three forms: gastrointestinal, cutaneous, and inhalation. Gastrointestinal anthrax happens after an animal or a person eats anthrax bacteria or spores—an extremely rare event in people. Cutaneous anthrax is far and away the most common type of naturally occurring anthrax.

Cutaneous anthrax begins when someone with an open wound accidentally contacts an infected animal (or, rarely, a person), and some of the bacteria move from the animal's skin onto the person's wounded skin. Most commonly, this has occurred in sheep shearers and cattlemen. After transfer, the anthrax bacteria begin to grow in the outer layers of the infected person's skin—the cutaneous portion of the skin. As the anthrax bacteria grow, they produce toxins that cause swelling, interfere with the host's normal protective responses, and cause inflammation.

The skin swells and some of it dies and turns black. This is the eschar or black necrotic lesion on the skin that is diagnostic for cutaneous anthrax. Usually, antibiotic treatment quickly cures cutaneous anthrax. Since the advent of modern antibiotics, few people have died from cutaneous anthrax.

Inhalation anthrax is a much more severe form of the disease. When anthrax spores land in human lungs, alveolar macrophages—white cells that help to defend our lungs—attack and swallow the bacteria. Then macrophages carry the anthrax spores to the nearest lymph nodes. Here, the spores germinate, and the real disease begins. Bacteria hatch from the spores and replicate until they burst their host cells. Then the bacterial toxins ooze into the surrounding tissues. This causes massive hemorrhages, swelling, death and necrosis of the cells of the lymph node, and a spreading inflammation.

Then the bacteria enter the bloodstream and, from there, they attack many other organs, including the brain and its membranes—the meninges. Death usually follows quickly, either from pneumonia and the lungs filling with fluid, high concentrations of bacteria in the blood and septic shock, or meningitis. As their brains and minds disintegrate, people drown in their own fluids. Sudden and horrible death.

Inhalation anthrax first appeared in the late nineteenth century. These first cases occurred in wool sorters in England. Apparently, during the handling of goat hair and alpaca fur from infected animals, anthrax and anthrax spores were blown into the air and inhaled by the workers, many of whom died.

And that's more or less the way it was until September of 2001—an occasional rare case, an occasional death.

Bacillus anthracis is a nearly ideal agent for bioterrorism. The bacterium grows rapidly in animals and humans until the host dies and the body splits open. Once in contact with air, the bacteria form spores— hard encapsulated forms of the bacteria that can survive for decades outside of any animal, waiting for another opportunity. For example, in the middle of World War II, the British used explosives to spread anthrax spores among a flock of sheep on the tiny Scottish island of Gruinard. The sheep died almost immediately, but the spores released that day were still viable thirty years later.

Anthrax bacteria and spores live in the soil in many areas around the world. It is relatively easy to grow anthrax from soil sources, and it is cheap—much cheaper than building bombs or hijacking airplanes. Most university libraries contain all the information needed to grow anthrax. And anthrax forms spores in the laboratory as well as in the real world, so spores are also readily available to those with the laboratory facilities needed for most microbiological work.

Generating "weapons-grade" anthrax requires a little more skill, but not much. These spores may be stored for years. In this form, the bacteria will survive almost anything, including mailing, and the fine powder will spread very nicely through the air to infect and kill as many as needed.

In 1993, a report from the U.S. Congressional Office of Technology Assessment estimated that the release of as little as 220 pounds of anthrax into the air over Washington, DC, would cause three million deaths—which is equivalent to the effects of a hydrogen bomb.

Only moist, high temperatures in combination with high pressure or radiation will kill anthrax spores. Very sturdy stuff.

Spores of *Bacillus anthracis* killed Robert Stevens and the others in the fall of 2001—weaponized anthrax spores packed in white envelopes and carried to their targets by the U.S. Postal System. Five letters filled with death, all dated 09-11-01, all mailed from Trenton, New Jersey. One sent to the *New York Post*, another to Senator Tom Daschle, a third to Senator Patrick Leahy, the fourth to Tom Brokaw of NBC, and one addressed to Robert Stevens at the *Sun*. Only four of the letters were recovered. The FBI never found the letter Stevens discarded.

For the most part, the aim of the terrorist was poor.

According to the UCLA School of Epidemiology:

The Daschle letter was opened in the sixth floor office at 9:45 am by an aide in the Senator's Hart Senate Office Building suite on October 15, 2001. It was believed to contain about 2 grams of powder comprised of 200 billion to 2 trillion spores. Based on nasal swabs, all 18 persons who were in the area of Daschle's sixth floor office tested positive for anthrax exposure, as did 7 of 25 (i.e. 28 percent) in the area of the Senator's fifth floor office (an open staircase connected the two offices).

The Leahy letter never arrived at his office. Instead an optical reader misread the hand-written 20510 ZIP code for the Capitol as 20520, which serves the State Department. As a result, the letter was routed to the State Department, where it arrived on October 15, infecting a State Department postal worker. Shortly thereafter, all mail was isolated and sealed in plastic bags for a later search.

Tales of great good fortune and fatal misfortune.

Similarly, aides handled the letters to the *New York Post* and to Tom Brokaw. But that's part of the point with terrorism—civilian casualties, lots of them. You never know where the axe will fall, but as long as it falls on someone, the terrorists have made their point. The mail or the seeming invulnerability of the people of the United States will never be the same again. Terror—death in an envelope, address unknown.

The Roots of Bioterrorism

For most of us, anthrax and the poisoned letters were the beginning, but 2001 was not the first year in history when bioterrorism changed the structure of power in the world. The first recorded use of a biological agent as a weapon occurred more than six hundred years ago in the Crimean city of Caffa (now Feodosiya, Ukraine). In 1343, a Mongol leader named Janibeg got into a brawl with some Christians, including some Italian merchants in the town of Talan. The cause of the fight is no longer clear. But the Italians quickly recognized the trouble that they were in and fled to nearby Caffa—a well-fortified city on the coast, a city full of merchants from all over the world, including many from Italy. The Italians imagined they would be safe there. They were wrong.

Janibeg and his horde followed. Apparently, the Italians had offended Janibeg very deeply, because for the next three years, Janibeg sat with his army of Mongols outside the city of Caffa, occasionally hurling stones with catapults and demanding the city's surrender. Because it lay directly on the coast of the Black Sea, Caffa still received food from the sea. So the people of Caffa waited, praying that Janibeg would tire of his siege. After three years, the people of Caffa got an answer to their prayer, but it wasn't the answer they were expecting. In 1346, an invisible foe attacked and decimated the Mongols waiting outside the walls. It was almost as though God had finally seen the plight of the people inside the walled city and rained down his vengeance on the evil hordes. It wasn't, though, an act of God. It was the black plague and the bacterium *Yersinia pestis*.

Eventually, there were mountains of dead among the Mongols, and Janibeg realized he was beaten. But Mongols were apparently very good at holding grudges. Before Janibeg left—beaten or not—he would have his revenge. As he was withdrawing with his remaining troops, Janibeg ordered thousands of the dead and diseased bodies and body parts catapulted into the city.

In Caffa that day, it rained plague-ridden pieces of men for hours. Whole rotting bodies, heads, arms, legless torsos, legs, fingers, and noses exploded like rotten melons against the cobbles of Caffa's streets. That night, the rats fed upon the rotting meat and consumed the plague. Fleas that fed upon the blood of those rats finished the destruction of the city of Caffa. And that ended most everything—except the plague, of course, which went on to devastate all of Europe and much of Asia.

Bioterrorism is an old and honored means of demoralizing and destroying your enemies. No military victory was won that day at Caffa, but revenge was taken, and to Janibeg, the defeat tasted as sweet as any victory he had ever achieved.

Almost seven hundred years later, the five deaths from anthrax were no less horrible. For bioterrorism, that is critical. Biowarfare is a somewhat different issue. In warfare, the purpose is to conquer and control. Usually, that is best accomplished with microbial agents that sicken many but kill few. Conquest of sick people is a lot easier than conquest of healthy people. And making use of a country's resources and generating vast numbers of slaves afterward is easier if you haven't murdered the entire population of the country you have just conquered. Conquering land is one thing; conquering land and people often offers a much richer reward.

Bioterrorism isn't like that. Bioterrorism isn't intended as a means of conquest. Bioterrorism, as its name implies, aims to terrorize, to instill a deep and lasting fear in people, and through that fear to manipulate those people's governments toward the goals of the terrorists. The microbiological agents of bioterrorism are meant to kill as many as possible as horribly as possible. Death, especially sudden, agonizing, and horrible death, instills fear in the living—deep and abiding fear. Imagine the suicide bombers in Israel, Afghanistan, and Iraq. The possibility that the person next to you on the bus or in the café might suddenly detonate himself or herself, shredding you and maybe a dozen others, lights the flames of fear in most of us.

But that fear is even greater when the terrorist is invisible, when death may come in a small cloud of white powder or in the gut of a flea. Fear of the unseeable and the unstoppable—that is terror.

Plague

> Ring around the rosy
> A pocketful of posies
> Ashes, ashes
> We all fall down.

Even without the rotting heads and the catapults, bubonic plague remains an attractive alternative to anthrax and the others as a means for making people's hearts tremble and fingers rot.

Most often, human plague begins inside the belly of flea. The flea sucks up the *Yersinia pestis* from rodents—commonly, rats—as the insect feeds on the blood of the living. Inside the flea, the bacterium plugs up the flea's esophagus.

Fleas make their living by sucking blood from their hosts. As they feed, fleas bite into their hosts' skin and chew through to the capillaries there. The host bleeds from those torn vessels and the fleas slurp and swallow the blood that pools in the bites. Inside of a feeding flea, the blood moves down the flea's esophagus and into its stomach, where the flea digests the blood and makes what it needs, usually more fleas.

But fleas infected with *Yersinia pestis* cannot swallow, because the bacterium has clogged their throats. As they slowly starve to death, the fleas become frantic to feed. They bite their hosts as often as they can in their hopeless drive for blood. When fleas bite, they inject a bit of saliva to prevent coagulation, to ease the flow of blood. *Yersinia pestis*, after it infects fleas and plugs up their fleas' esophagi, migrate to the fleas' salivary glands. Then, each time the fleas bite, they squirt plague along with saliva into the wound they have opened.

Eventually, infected fleas die of starvation. But by then they have served their purpose—syringes full of plague injecting as many bacteria as possible before the fleas die. Once injected, the *Yersinia pestis* no longer needs the fleas. Now it moves quickly through its host's lymphatics into lymph nodes—little filtering stations in the neck, under the arms, in the groin and the abdomen. Inside lymph nodes, the bacteria multiply and the immune system struggles to control them. But the bacteria begin to inject proteins into the host's macrophages. The proteins the plague bacteria inject take control of the macrophages and prevent the host's cells from attacking the invading microorganisms.

During the battle, the lymph nodes swell enormously and painfully. These are the so-called buboes of bubonic plague.

A horde of new bacteria spills out of the lymph nodes and through the thoracic duct into the bloodstream. From there, *Yersinia pestis* goes everywhere: the kidneys, the spleen, even the lungs.

At this point in the infection, blood begins to seep from people's veins under their skin, causing rose-colored bruises.

Ring around the rosy.

Temperature rises. There are fever and chills, vomiting, diarrhea, black and tarry stools. Blood flow to the extremities fails. The death of tissue—black necrosis—follows in the nose, the penis, the toes, and the fingers. Pieces of people begin falling off. The stench is horrid. People often brought flowers to cover the smell.

A pocket full of posies.

In the lungs of infected people, plague multiplies and causes massive inflammation. Mucus oozes from the cells lining the lungs. An infected person begins to cough, hacking up great clots of mucus and millions upon millions of bacteria. All of it spews into the air with every cough. Anyone nearby may inhale it. Those who do develop an even deadlier form of plague called pneumonic (for lung) plague. There is no more need for fleas, and there is no need now for the plague to spread through the blood to reach the lungs. It takes a week or so for symptoms to show in a person with bubonic plague and several days more to die from the infection. People with pneumonic plague develop symptoms within a day, and if untreated, 90 percent of these people will die within a few days.

It begins in the lungs and quickly multiplies—inflammation flows along with mucus, coughing, and more infections. Pneumonic plague spreads from person to person, and it spreads like water from a burst dam. Just as it did with the fleas, *Yersinia pestis* has quickly converted a human being into a machine for making more bacteria and spewing them into others.

Ashes, ashes.

Death—ashes to ashes. Or, perhaps, achoo, achoo from the sneezes of those with pneumonic plague.

Without treatment, 50 to 90 percent of infected people will die. Even with the best treatment, somewhere around one in six will die.

We all fall down.

Imagine, as Janibeg did, the effect of watching the sky above you fill with rotting human heads full of plague, imagine the sound as they exploded onto the stone cobbles, imagine the agonizing deaths that followed. Envision that, and you have seen the flame at the dark heart of bioterrorism.

The Future of Bioterrorism

Nuclear weapons are very showy, like Ferraris. But also like Ferraris, nuclear weapons are very expensive. You need aluminum tubes and centrifuges and leftover radioactive waste, and you need a delivery system. All very expensive and all very closely monitored.

None of the stuff you need to grow anthrax or plague or Ebola virus or Q fever or tularemia or Marburg virus is controlled, closely monitored, or expensive. Even the agents themselves are fairly simple to come by, especially if you have a contact at a hospital in sub-Saharan Africa. Generally, contaminated bodies or body parts can be had for free or for nominal bribes. After that, you do need a laboratory, but it can be fairly rudimentary. A biosafety cabinet is important if don't wish to kill yourself or those working for you—a few thousand dollars. You need some electricity, two or three small centrifuges, some broth, and a few animals—monkeys or rats, depending on what you want to grow. Some syringes, some needles. A couple thousand dollars more. And if your supporters are as dedicated as many Palestinian or Iraqi terrorists, then you already have your delivery system.

The U.S. Department of Health and Human Services, which has oversight responsibility for the health of the American people, has compiled a list of the agents they consider to have potential use in bioterrorist attacks. This is the "Select List" of agents. The list contains eighty-three biological agents and toxins. Thirty-two of these affect primarily or solely humans; twenty affect both domestic livestock and humans.

Toxins are substances produced by living things that could be used to kill people, things like ricin.

Ricin is a protein found in castor beans. Castor beans are processed throughout the world to make castor oil, which is widely used as a brake and hydraulic fluid. After the oil is extracted from the beans, the leftover mash is 5 to 10 percent ricin. Isolating the toxin from the mash requires little money and an undergraduate college student's knowledge of chem-
When ricin gets inside of the body's cells, it shuts down all protein

synthesis. When that happens in the heart or lungs, people die very quickly. Only about 200 micrograms are needed to kill a 165-pound man (a microgram is one-millionth of a gram). Pennies' worth of ricin. That has some appeal to those who wish revenge or to instill terror.

Between 1991 and 2003, four investigated incidents in the United States involved the use of ricin. In 1991 in Minnesota, police arrested four members of the Patriots Council for plotting to kill a United States marshal by using ricin. The Patriots Council was founded on antitax and antigovernment principles, and more than once the group had advocated the overthrow of the U.S. government. The Patriots' plan was to mix the ricin with a solvent, dimethyl sulfoxide (DMSO), and rub it on the door handle of the marshal's patrol car. DMSO will carry many chemicals through human skin. The ricin involved had been whipped up in someone's basement.

In 1995, Canadian customs officials stopped a man traveling from Canada to North Carolina. When they searched the man, they found $98,000, several guns, and a "container of white powder." Nothing from the Centers for Disease Control (CDC), of course, would indicate what the traveler planned to do with the guns or the white powder that turned out to be ricin.

Then, in 1997, after a man shot his stepson in the face, police uncovered a basement laboratory filled with ricin and nicotine sulfate (a strong irritant and nerve toxin). Again, no indication was given of what the man might have been planning before he decided to take time off and shoot his stepson in the face.

Finally, in October 2003, ricin showed up in an envelope at a mail-processing facility in Greenville, South Carolina. The letter was protesting a proposed federal limit on the number of hours long-haul truckers could spend on the road. The letter was signed "Fallen Angel." The next month, another ricin-laced letter showed up at a White House mail-processing center. This letter was also signed "Fallen Angel."

Internationally, ricin has turned up in Great Britain on at least three occasions. On September 11, 1978, Bulgarian dissident Georgi Markov left his office at Bush House in London and walked across the Waterloo

Bridge to catch his train home. Markov was a communist defector working for the BBC World Service. As he waited at his stop, he felt a sharp pain in his calf. When he turned around, he saw a man picking up an umbrella. The man with the umbrella apologized, but four days later Markov was dead. The umbrella was, in fact, a type of air gun constructed for a single purpose—to discharge a tiny pellet holding 200 micrograms of ricin into Markov's leg. The plot was discovered only because the pellet hadn't dissolved as planned. It appears, though no one has verified it, that the Soviet KGB was responsible.

In December 2002, a ricin laboratory in Manchester was raided, and six terrorist suspects were arrested. One of the arrestees was a twenty-seven-year-old chemist. And in January 2003, ricin was again found in London during arrests that eventually led to the investigation of what appeared to be a Chechen-separatist plan to poison the Russian embassy with ricin.

Toxins, like biological agents themselves, are cheap and easy to deliver. For that reason, if none other, they will remain among the top on the government's Select List.

Also on the Select List are fifteen agents that sicken and kill both livestock and humans—like *Bacillus anthracis*—and five toxins that affect both animals and humans—like *Botulinum* toxin. *Botulinum* toxin blocks the function of nerves, causing muscles to relax suddenly. It is one of the most lethal toxins known, and it would take only about 75 nanograms to kill a 165-pound human. That's about two thousand times more toxic than ricin. Three ounces of *Botulinum* toxin would be enough to kill a billion people.

Botulinum toxin comes from *Clostridium botulinum*, another bacterium that grows in soils throughout the world. Producing enough toxin to kill a few thousand people would cost pennies.

The other use of *Botulinum* neurotoxin is, of course, for cosmetic purposes (Botox).

Also on the Select List are another twenty-four agents that affect only animals, yet could have a severe impact on agriculture and the lives of many, many human beings. Beyond that, the list contains eleven agents that attack domestic plants and destroy crops.

A person with even minimal scientific knowledge could prepare most of these eighty-three agents and toxins, each of which is capable of instilling terror and disrupting millions of lives. And, as I said, delivery systems are cheap. The Select List doesn't even consider what could be done by genetically altering normally harmless microorganisms—a thing that gets easier to do every day. Biological terror is a very attractive alternative to conventional weapons of mass destruction.

These are facts that have not escaped the U.S. government's sometimes watchful eye. In recognition of the dangers of bioterrorism, the United States is spending $531 million to construct new research laboratories to study the agents of bioterrorism, $591.9 million to fund research on these agents, and $194.3 million for development of drugs, vaccines, and diagnostics. The National Institute for Allergy and Infectious Diseases has also added $18 million to the National Center for Research Resources for "biodefense-related construction projects."

Forty million dollars of these funds have been awarded to Colorado State University in Fort Collins. These funds will be used to establish a regional center of research excellence. Construction has already begun on sophisticated laboratory facilities to handle a variety of infectious agents, including tularemia, Q fever, plague, and so on. The biosafety level (BSL) rating of a laboratory determines the danger of the organisms that the lab can safely handle. The highest level is BSL-4. As of 2004, there were only four BSL-4 facilities operating in the United States: one at the Centers for Disease Control and Prevention in Atlanta; one at the United States Army Medical Research Institute for Infectious Diseases at Fort Detrick, in Frederick, Maryland; one at the Southwest Foundation for Biomedical Research in San Antonio; and one at the University of Texas in Galveston. The regional center of research excellence at Colorado State University will rely on BSL-3 laboratories for its investigations.

Dr. Ralph Smith in the Department of Microbiology, Immunology, and Pathology at Colorado State University oversees the construction of these laboratories. "I'm reassured by the government's commitment to this type of research," he told me. "For both newly emerging or reemerging infectious diseases as well as for the efforts of bioterrorists, we will be much better prepared. These efforts involve a consortium of universities and the best scientists. At CSU, we'll focus on zoonotic diseases [those transmitted from animals to people]. We are the only center in the United States devoted to zoonotic diseases. But that's where our true expertise lies, especially with the veterinary school here and our historical roots in agriculture. And we have the Vector-Borne Division of the Centers for Disease Control here as well. That should make for a very effective and focused effort against these agents."

I asked what worried Dr. Smith most among potential bioterrorist threats.

"Genetic modification of existing agents," he said. "Something simple, something we've known about forever, but with a new gene, a new ability to cause disease. That could be catastrophic."

Genetic engineering was supposed to be humankind's savior.

The more questions I ask of people like Dr. Smith, the less I wish to go on asking these questions.

Our relationship with infectious agents is billions of years old. For most of that time, we have managed to hold our own with the microscopic universe. But now it seems that the things that kept us safe all those years are no longer enough. Once humans start intentionally infecting other humans, the old rules just don't apply anymore.

In my opinion, it appears that Dr. Smith is right—the U.S. government is mounting a tremendous effort to limit the effects of bioterrorists. Let's hope it will be enough. We have many people to thank for that, among them certainly people like Dr. Smith and the other scientists who have worked to put all of this in place and will work toward better understanding of these agents and how to protect all of us. But first and foremost on our list of those who deserve our respect should be Robert Stevens, Thomas Morris Jr., Joseph Curseen Jr., Kathy Nguyen, and Ottie Lundgren—the people who died in 2001 to warn the rest of us about what was out there.

13 ⁘

Eating Your Brains Out:
Mad Cow Disease

An Acquired Taste

The drums rolled slowly beneath the old palms and the air steamed with the froth of the sea. It was to be a great feast. Both tribes had reason to celebrate. Roasted meats and fruits, dried dates, and fruit bats wrapped in banana leaves, smoke from the fires, and sweat from the men and women filled their heads with wishes and their mouths with saliva.

The men had smeared brilliant blue and fiery red or white paint across their faces. One man bore the brittle white stripes of a human skeleton, another wore the face of a jaguar, all sported feathers. The women had bathed in sweet oils and wrapped themselves in bright skirts, their hair pulled back.

Talk was difficult, not just because of the din of the drums but also because the tribes shared few words. But the importance of the meeting and all that it portended were obvious. The Fore were a great people, but this other tribe had things the Fore had never seen or heard—words like "birds" and meats the Fore had never tasted.

They sat with one another, and they ate the salted flesh and the sweet fruits until no one was hungry. Then they drank the juice from fermented bananas until no one was sober.

Finally, in their inebriation, someone asked about the deeply browned and piglike meat that had been served in such abundance. Words failed, but ideas and motions didn't. What the Fore had eaten with such hunger, what had been the most sumptuous of all the animals roasted that evening, was human. This new tribe had brought to the feast thighs and breasts and wrists and calves carved from men and women.

The Fore felt honored and were greatly impressed.

That's how it probably began, and though I have taken some liberties with the specifics of the event, the feast itself does seem to have started it all.

The Fore were so impressed that after that night in 1915, for the next fifty or sixty years, the Fore ate themselves. Ate one another in formal rituals—not the living, just the dead among them. After a man or woman of the Fore died, after the death was celebrated, and after the body was allowed to putrefy for three or four days, the living roasted the dead. And then, because they believed the dead longed for it, they ate the body. But not haphazardly—there were rules. First, the deceased's mother's brother's wife got to eat the brain. And then the mother and the spouse and the sisters consumed the rest of the dead body in a specified order. The men, for some reason, believed that eating their ancestors would lessen their skills and endurance as hunters, so they skipped the meal.

Those meals would haunt the Fore in ways they never imagined.

At first nothing changed. The people ate the flesh of one another, and the world seemed just as bright, brighter, even. The birds still sang, the moon still swung low over the sea, but now there was food in greater abundance than ever before. Nearly ideal.

But one day, all of that changed. Five or ten years after they began to feed upon their dead ancestors, a new evil was visited upon the Fore. Young women began to dance, but not as any had ever danced before. Now they twitched with an unholy fire. Their heads flew from side to side and they trembled like leaves in a fierce wind. None among the tribe had ever seen anything like it. Not long after the women began to dance, they lost all sense, all memory, and, invariably, they died.

For the Fore, it seemed that this terrible affliction overtook their women when the ghost wind, or *zona*, blew cold from the north. And soon it seemed to all that that wind was conjured by witches among them. The Fore called this terrible dance to the death *kuru*, meaning "to shiver and tremble." Some of the people created intricate rituals to try to stave off this witchcraft; others tried to inflict the disease upon their enemies.

Those who died of leprosy or with diarrhea, they buried. But those who died dancing, they ate. And there were always more who danced.

None among the Fore ever imagined that the witchcraft involved came to them not from evil gods or fierce witches but from the dead themselves.

In 1954, an Australian patrol officer, J. R. McArthur, was passing his rounds in the highlands of Papua New Guinea, among the Fore. A young woman approached McArthur as he worked, and he noticed that she had a severe tremor causing her to shake violently, and her head jerked horribly from side to side. She passed him by without a word, but he would not forget her. When McArthur asked about her, the people told him that this was not unusual and that some witch had cursed this girl. McArthur suspected otherwise. By 1954, Papua New Guinea was a trust territory of Australia, so the control officer mentioned what he had seen among the Fore in his report to the Australian government.

A year later, medical officer Vincent Vigas began his investigations of the Fore. The Fore mistrusted him completely. Even so, Vigas was able to perform most of the routine sorts of diagnostics on some of the affected women. Nothing showed up—no bacteria, viruses, fungi, or parasites were infecting these women. Vigas was baffled.

In 1956, D. Carleton Gajdusek joined Vigas among the Fore. Gajdusek had just finished his research at the Walter and Eliza Hall Institute in Melbourne, Australia, and had heard of the disease described by Vigas, the disease the natives called kuru.

At first, nothing Gajdusek tried turned up anything either. Kuru appeared to be a disease without a cause. But Gajdusek was certain he was missing something.

Some of Gajdusek's findings attracted the attention of others working in similar fields. In 1959, a veterinarian named W. J. Hallow writing in the British medical journal *Lancet* serendipitously mentioned the striking similarities he had noticed between a disease of sheep, called scrapie, and what he had heard about the disease kuru.

At autopsy, the brains of people with kuru showed defects that looked like those seen in sheep with scrapie. He noticed, too, that sheep with scrapie stagger and tremble, much like the women with kuru. As they stagger, the sheep scrape off their wool against fences and trees, thus the name scrapie. But scrapie had been around for as long as people had been herding sheep. The cause was still mostly unknown, but the disease was centuries old. Using experimental animals, scientists had long ago shown that scrapie was transmissible, that is, infectious in sheep. Normal healthy sheep developed scrapie about two years after injection with brains from infected sheep.

With that information foremost in his mind, Gajdusek returned to New Guinea and collected brain tissue from several patients who had died from kuru and sent the specimens off to the National Institutes of Health in the United States. There, scientists inoculated chimpanzees with Gajdusek's samples. In 1965, the first chimps developed symptoms that looked very much like the symptoms of kuru.

About the same time, other researchers found similar results when they injected nerve tissue from human patients with Creutzfeldt-Jakob disease (CJD) into chimpanzees. CJD is another neurological disorder of humans that causes symptoms like scrapie. CJD, also called sporadic CJD, apparently occurs spontaneously in about one in a million humans. The symptoms, like those of kuru, involve a progressive dementia along with muscle tics and twitches, loss of motor function, and finally, and invariably, death.

A year or so after chimps were injected with brain tissue from CJD patients, the chimps developed diseases that looked like CJD and kuru.

While it was still unclear what caused any one of these three diseases, it appeared they all had similar roots. At autopsy or necropsy, the brains from humans with CJD or kuru and sheep with scrapie all had millions of tiny holes in their brains where neurons should have been. Instead of healthy organs, their brains looked like sponges. And like scrapie, CJD had been around for decades.

The researchers began to suspect a relationship among all three diseases.

Prions and Disease

Together, kuru, CJD, and scrapie became known as transmissible spongiform encephalopathies or TSEs. Transmissible, because the disease could be transferred from animal to animal. Spongiform, because the disease turned people's brains and sheep's brains into sponges. And encephalopathies because *encephalo* = "brain," *pathy* = "disease." Something was eating up large chunks of the brains inside people and sheep. And it was something unlike any other transmissible agent known.

In 1982, Stanley Prusiner finally developed a laboratory assay system that allowed him to find out what that something was—the thing that not so slowly destroyed people's and other animals' minds and brains. Using Syrian hamsters, Prusiner discovered that what was turning brains to mush wasn't a virus or a bacterium or a fungus or a parasite. Instead, it appeared to be a simple protein—no genes, no reproduction, no life force, no evolution, no motivation, just a protein with a wicked habit of twisting people's spines and eating out their brains.

No scientist, even in his or her worst nightmares, had ever imagined anything like that. At least with viruses and bacteria and fungi and parasites we rationalize their actions by assigning reproductive motives. Biological drives toward reproduction have been evolving over millennia. They are powerful things that all of us need for survival. But a protein? No DNA?

Prusiner named these proteins prions. The part of a virus that does the real work of viruses, the part that causes disease, we call the virion— *vir* for "virus." Prusiner followed this pattern with his protein to create prion.

He provided very strong evidence that it was prions in the brains of the dead that caused kuru in the living who ate their ancestors. In fact, after the discovery of prions and TSEs, anthropologists and public health workers convinced the Fore to stop eating their ancestors. Because it was a recently developed tradition, and because there was little religious significance attached to their cannibalism, the Fore gave it up. Since then, kuru has disappeared from Papua New Guinea.

Besides kuru, prions appear to cause diseases in many species of mammals—scrapie in sheep and goats; transmissible encephalopathies in mink; feline spongiform encephalopathies in cats; CJD and its variant vCJD; fatal familial insomnia (FFI) and Gerstmann-Straussler-Scheinker disease (GSS) in humans; bovine spongiform encephalopathy (BSE) or mad cow disease in cattle; and chronic wasting disease (CWD) in deer, elk, and moose.

With the exception of CWD and the genetic forms of prion diseases (GSS and FFI), transmission of spongiform encephalopathies occurs most often when healthy animals eat the brains or spinal cords of infected animals.

After that, things get murky.

Clearly, prions survive the hydrochloric acid and enzymes in the digestive tract, then somehow travel from the gut to the brain, where they cause disease.

But here's the most amazing part about these diseases: We all have prion proteins in us, all of the time. If we didn't, we couldn't get TSEs. But the prion proteins inside most of us are slightly different from the prion proteins inside people with TSEs.

To distinguish them, scientists have given these two proteins different names—PrP^c is the normal or "cellular" form of the prion protein, while PrP^{sc} is the pathogenic or disease-causing form of the protein.

The "sc" comes from scrapie, the prototypical TSE. Interestingly, the only difference between the PrP^c and PrP^{sc} is the shape of the two proteins; chemically, the two molecules are identical.

In simple terms, PrP^c is curlier than PrP^{sc}. The transition from PrP^c to PrP^{sc} has been described as "rather like turning a chiffon curtain into a Venetian blind." And most important, for reasons no one quite yet understands, PrP^{sc} will kill you but PrP^c will not.

The PrP^{sc} protein appears to cause disease in a most remarkable way. Inside of animals' bodies, PrP^{sc} changes PrP^c into more PrP^{sc}. That is, the "bad" form of the protein has the ability to turn the "good" form of the protein into more of the bad form.

So that after a while—sometimes as much as ten or more years—much of an infected person's PrP^c becomes PrP^{sc}. Some of that PrP^{sc} accumulates as a sticky goo, called amyloid, in a person or animal's brain. And somehow that causes the destruction and removal of neurons and holes in people's minds.

As brains become sponges, cognitive and motor functions fail. At first, thinking slows, people have difficulty concentrating, and they begin to make poor judgments. Memories slip away. Personalities and behaviors change. Then dementia takes over and self-neglect and apathy follow. Muscle spasms cause these people to twitch mercilessly. Seizures are common. At the end, all mental and motor functions are lost. People become bedridden and lapse into comas, and then fatal infections, like pneumonias, finish these lives.

A terrible disease.

Even worse, all normal methods of sterilization have no effect on prions—not cooking, not boiling, not traditional pressure-steam sterilization, not gas or chemical sterilization. Prions are remarkably stable little proteins.

Fortunately for human beings, naturally occurring prion diseases are rare in people. Only about one person in a million will develop CJD in his or her lifetime. For that reason, most of us worry little about CJD. But that is changing.

Mad Sheep, Mad Cows

Even though people had known about scrapie since the eighteenth century, and even though people had known since 1936 that scrapie was a transmissible disease, cattle were fed sheep until 1997.

Somewhere along the line, it occurred to sheep ranchers that much of each sheep was going to waste. People would eat only a fraction, mostly just the muscle of the animals, and the rest was thrown out. The carcasses still contained bits of meat, intestine, lungs, hearts, spleens, thymuses, bones and bone marrow, and blood, and brains, and spinal cords—nearly 50 percent of each animal in total. These carcasses were a great source of proteins and minerals, if only they could be packaged appealingly. So sheep ranchers called in the animal feed industry people, and together they came up with a plan—rendering.

Once slaughterhouses removed the parts that people regularly eat, the carcasses were pulverized in machine-driven grinders and boiled in vats. This produced a slurry of dissolved protein under a layer of fat. Workers then skimmed off the fat and dried the remainder into a meat-and-bonemeal product that they sold as feed for domestic livestock, zoo animals, laboratory animals, and pets.

Sheep ranchers found a new source of income, and sheep found their way into cattle.

About the same time, cattle ranchers realized that most of their cows were going to waste. So they followed suit.

Then cattle found their way into cattle.

And everyone was pretty happy, save the cattle and sheep, until about 1986. That year in Great Britain, a cow or two went mad. At necropsy, the cows' brains looked like lace, much like the brains of sheep with scrapie.

A ripple crossed the pond, but everyone looked the other way. The ripple, though, built as it rolled across the water. By 1988, nearly three thousand U.K. cows went mad, and everyone suspected the cattle feed, which included ground-up brains and spinal cords from sheep and cattle. In 1988, the British government banned some animal products in cattle feed, but the numbers of mad cows continued to increase. In 1991, more than twelve thousand cattle became ill. That same year, the government specifically banned "specified bovine offal"—brain, spinal cord, thymus, tonsils, spleen, and intestines from cattle greater than six months of age. Still the numbers of mad cows rose. In 1993, more than fifteen thousand cows developed what is now known as bovine spongiform encephalopathy or BSE. Then things began to get better for cattle.

Not so for people.

It can take decades for people infected with prions to begin to show symptoms. So there was a lag phase, a time while everyone held their collective breath and waited to see what would happen with people.

Fears worsened when zoo ungulates (hoofed animals, such as antelopes, camels, rhinoceroses) in the United Kingdom developed spongiform encephalopathies. Investigators found that all of these animals had eaten diets supplemented with animal by-products and bonemeal. Then domestic and wild cats started going mad. The domestic cats had also eaten dietary supplements that contained meat by-products and bonemeal. The wild cats, it turned out, had eaten uncooked meats that included animal heads and spines. Clouds were building on the horizon.

Still people waited. Scrapie, they said, is not a human pathogen. And they were right. Nobody ever seemed to have gotten scrapie from sheep. But there was lots of evidence that after scrapie passed through other species—like mice or hamsters—it changed, and those changes led to new infections in species that had been resistant. And still people waited.

For ten years, nothing seemed to change. The incidence of CJD, even among high-risk groups of people, stayed about the same as it had for decades.

But the storm finally broke in 1995. By this time, more than two hundred thousand cattle had died from BSE and 4.5 million more had been destroyed to try and slow the epidemic. And then, almost ten years after the first mad cow, between May and October the CJD surveillance unit was informed of three cases of CJD in people ages sixteen, nineteen, and twenty-nine.

Less than 10 percent of cases of sporadic (or regular) CJD affected people that young. Something had gone wrong.

By December of 1995, the surveillance unit was handling ten more cases of CJD. Two of these were like the first three the unit had seen — new, different. Two more probable cases of this CJD turned up in January of 1996, three more in February, twelve more in March. "Probable" patients become certain patients only at autopsy.

But already it was clear that this was CJD unlike any seen before. The people who discovered the new TSE called it vCJD or variant CJD. This TSE killed much younger people—an average of twenty-eight years of age for vCJD victims versus sixty-eight years of age for CJD. vCJD caused more prominent mental and behavioral changes early in the course of the disease. And vCJD prions accumulated earlier and in larger numbers in amyloid deposits in brains and lymphoid tissues. vCJD was a clearly distinct form of the disease.

The potential link between vCJD and the BSE epidemic in cattle was obvious to nearly everyone.

In 1996, the British government banned feeding meat and bonemeal from any mammal to any other mammal.

Of course, no one could experiment with people to prove that vCJD came from cattle with BSE. So to some extent, whether or not people actually got vCJD from cattle remains moot. But the vCJD outbreak followed the pattern seen with the cattle, plus a lag phase of a few years. The peak year for mad cow was 1992–1993; the peak year for vCJD in humans, so far, was 2000. Curiously, there was also a notable increase in sporadic CJD that began in about 1992 and peaked in 1998–1999.

By 2003, 153 people had died of vCJD: 143 in Great Britain, 6 in France, and one person each in Canada, Ireland, Italy, and the United States. The people who died in Canada, Ireland, and the United States had resided in Great Britain during the BSE outbreak. And essentially all of the 153 who died had multiple-year exposures in the U.K. between 1980 and 1996, peak years for the outbreak of BSE there.

Many scientists had believed that prions would not jump between species. vCJD seems to have proved them wrong. Recent research studies have shown that BSE can also be transferred to other species under the right conditions. However, to date, there is no evidence that scrapie or chronic wasting disease of deer and elk (another TSE infecting animals we humans regularly eat) can infect humans. Mad cows appear unique with respect to human risk, even though it seems that cows got BSE from sheep scrapie, which may also be the origin of chronic wasting disease.

There is, however, another interesting theory put forward recently about the origin of mad cow disease, one that makes its infectivity in humans even more understandable.

Most Hindus believe that the dead should take their final voyage down a river, if possible, the Ganges. Traditionally, families burned the bodies of their dead. But now, very few can afford the firewood needed. It is now customary simply to smoke the pelvis of a woman or the upper body of a man. After that, most of the bodies are cast into the river, whole. In Varanasai, a holy city near the Ganges, more than forty thousand funerals take place each year. From a single six-mile stretch of the river, volunteers removed sixty human bodies in two days in 2004.

In India and Pakistan, many workers make a living gathering bones and carcasses from the farmlands and the rivers. Human remains are often gathered along with those of cattle.

In the 1960s, Britain imported hundreds of thousands of tons of whole bones, crushed bones, and carcasses for use in making fertilizer and animal feed. Almost half of this grisly product came from Bangladesh, India, or Pakistan. Though some of the bones and tissues were intended as fertilizer, probably some of what arrived in Great Britain found its way into cattle feed.

In September of 2005, writing in the British Journal *Lancet*, Alan and Nancy Colchester proposed that BSE came to cattle from human remains that ended up in cattle feed. All of the information described above led them to believe that human bones and soft tissues collected along with the remains of cattle in India and Pakistan might well have been enough to start Britain's epidemic of mad cow disease. That would, of course, explain why the disease moved so easily from cattle to humans—it was only going back to where it had come from. But at this point, all of it is still conjecture. We may never know with any confidence where or how the prion came to British cattle.

Why the BSE prion but not the scrapie prion may infect humans isn't clear, but it could be that the PrP^c of cattle provided the link. Even though PrP^{sc} from sheep does not seem to affect humans, it could create altered cellular prions in cattle. And these new PrP^{sc} prions might be able to infect humans and cause vCJD.

Regardless, how could a person get vCJD from eating beef?

Obviously, a person could eat the brains or spinal cords of cattle. Some people do. Once BSE is in full swing inside of a steer, the brain is loaded with PrP^{sc}. So even a well-cooked brain or cord could provide a considerable helping of bovine PrP^{sc}. But few of us ever did eat cow brains or spinal cords, even in Britain. So what about the rest of those cases of vCJD?

Certain cuts of meat—like T-bone steaks—are butchered from the short loin, very near the spine. It would not be difficult to contaminate T-bone steaks with spinal cord tissue, and again, cooking would be no help.

Still another way the disease might move from cattle to humans has to do with the way beef cattle are slaughtered. A common method is to use a captive-bolt stunner. This is a gunlike device that is pneumatic or powered by an explosive cartridge. Both types of guns fire a steel bolt that is attached to the gun so it can be retrieved and used over and over (captive bolt). Some captive-bolt stunners also inject air into the crania of cattle. But the pneumatic or explosive guns, or those that inject air, don't kill the animals. That is not the intent.

The cattle move into a chute one at a time and are restrained. The stun gun is then placed against the cow or steer's head and fired. The stunned, but still living animal is hoisted, the jugular vein is severed, and the animal bleeds to death. The stunning is done to comply with federal laws for humane slaughter of livestock. These laws require that "animals be rendered insensible to pain before being hackled, hoisted, thrown, cast, or cut (unless they are slaughtered and handled in connection with slaughter in accordance with certain specified religious ritual requirements)."

As you can imagine, the captive bolt does considerable damage to the animals' brains when it strikes their skulls. More interestingly, from the mad cow perspective, the bolt ruptures the brains of the animals and sends brain tissue into their blood. That has considerable potential to transfer prions to the lungs, liver, heart, and even muscle.

After stunning, animals remain alive for several minutes before bleeding to death. As long as the heart continues to beat, pieces of brain may circulate with the blood. The potential for the spread of prions in infected cattle is obvious.

Finally, beyond all of that, in humans with CJD, deer with CWD, and mice with TSE, scientists have found that even without head trauma, prions are present in many places outside the central nervous system, including muscle tissue. Of course the muscles are the parts of cattle that most of us eat—steaks, burgers, roasts, brisket, etc. So it is possible, even before slaughter, that PrP^{sc} could be present in any cut of beef from mad cows.

In 2001, the U.S. Department of Agriculture commissioned a study managed by the Harvard Institute for Risk Assessment, in the Harvard School of Public Health and in the Center for Computational Epidemiology in the College of Veterinary Medicine at the Tuskegee Institute. The purpose of the study was to investigate the potential of BSE to become a major animal health problem or substantially contaminate the human food supply in the United States. The committee assembled for that risk assessment concluded that the risk of an outbreak of BSE in the United States was minimal.

But, they added, "In short, the U.S. appears very resistant to a BSE challenge, primarily because of the FDA feed ban, which greatly reduces the chance that an infected animal would infect other animals. However, the effectiveness of the feed ban is somewhat uncertain because compliance rates are not precisely known."

In fact, compliance rates are pretty much totally unknown. That's bound to have some consequences for these predictions.

And beyond the risk of spreading BSE to more animals by contamination of feed with animal protein, what about the risks to human beings?

The Harvard-Tuskegee report determined that there are two main risks to the human population. Eating brains and spinal cord is the obvious one, but few people intentionally do that. Many more of us may unintentionally do that.

You might imagine that if you stick with steaks and burgers, you won't be eating cattle brains or spinal cords. You might be wrong.

The Harvard-Tuskegee report noted that there was another major risk for contamination of other meats due to a process called advanced meat recovery.

According to the American Meat Institute: "Advanced meat recovery (AMR) is a technology in place in some meat plants that uses a special machine to rub meat off beef and pork bones that can't be hand trimmed. Just as fruit processors use machines to remove fruit from peels thoroughly and efficiently, meat companies use similar equipment to remove meat from some hard to trim bones."

Most of this meat comes from back and neck bones. Because of the potential for contamination with nerve tissue, the USDA requires that processing plants remove the spinal cord from the spine before AMR. "While [according to the American Meat Institute] spinal cords are safe they are not meat and are not permitted in meat products."

It is the process by which the spinal cord is removed that leads to the problems described in the Harvard-Tuskegee report: "Our analysis indicates that the most important means by which low risk tissue can become contaminated is the use of advanced meat recovery (AMR) technology, which can leave spinal cord or dorsal root ganglia (DRG) [also nerve tissue] in the recovered meat. Our analysis further indicates that mis-splitting of the spinal column and the resulting incomplete removal of the spinal cord is largely responsible for contamination of AMR meat. In addition, we assume that even in the absence of mis-splitting, some amount of DRG is extracted whenever vertebrae are processed by AMR."

By law, AMR meat can be called "meat" and can be included in any meat product. But most of it ends up in ground meat, meatballs, and taco fillings.

The Harvard-Tuskegee risk assessment does not address the studies that have found CJD, CWD, and mouse prions in the muscles of untraumatized animals. Also, in response to the USDA's request for a peer review of the Harvard-Tuskegee report, a group of scientists at RTI International in Research Triangle, North Carolina, produced 132 pages of criticism of the Harvard-Tuskegee risk assessment. The scientists involved with the original report then revised some of their statements and conclusions. However, the scientists at Harvard and Tuskegee remained convinced of the validity of their original assessment—there is little risk to cattle or humans in the United States from BSE.

Of course, the ultimate validity of the report relies heavily on assumptions about the contents of cattle feed. In the end, the potential for contamination of the human food supply is directly proportional to the number of mad cows that could be brought to slaughter and processed for human consumption. And the risk for the spread of BSE among cattle is directly proportional to the likelihood that cattle will be fed tissues from infected sheep or cattle. Because of that, animal products in cattle feed remain the prime concern.

On April 16, 1996, the Oprah Winfrey show's guests were Gary Weber, spokesman for the National Cattlemen's Beef Association, and Howard Lyman, a former beef rancher turned vegetarian.

When Winfrey asked Weber if we were feeding cattle to cattle in the United States, he replied that, in fact, we were. The audience was a little taken aback. So Weber quickly pointed out that cows do drink milk, an animal product.

Winfrey asked Lyman why he was now a vegetarian, and he responded that it was because of what he knew about what was happening out there with cattle.

Lyman explained that even though voluntary bans on feeding animal by-products to cows had been in place for years, at least a quarter of the rendering plants that the government had checked were ignoring the ban—even though the British by this time had had mandatory bans in place and enforced for nearly a decade.

Someone then asked Lyman why we fed animals to animals. He explained that about half the steer (by weight) remains after the meat we normally eat has been removed. The remains, he said, we can either pay to put in the dump or we can feed it (along with road kill and euthanized [meaning, in my mind, sick] animals) to our livestock.

Apparently, at the time, it seemed more sensible and economical to many people to feed these pieces of dead cattle and other animals to live cattle and other domestic animals rather than simply bury them. But some other people, including Oprah, were having second thoughts about hamburgers.

In 1988, Great Britain instituted a ban on feeding sheep or cattle protein to cattle. In 1989, the U.S. rendering industry imposed a voluntary ban on the use of adult sheep tissue in ruminant feed (ruminants here means primarily sheep and cattle). An investigation in 1992 found many rendering plants, particularly the smaller ones, were not in compliance with the ban. The United States did not institute mandatory bans on feeding animal parts to cattle until 1997, more than ten years after the first cases of mad cow disease were diagnosed in Great Britain.

The first mad cow discovered in the United States showed up in 2003. Though discovered in the state of Washington, this steer apparently came from a herd in Canada. So the USDA immediately placed a ban on importation of Canadian beef, effectively closing our northern border to cattle, and declared U.S. cattle herds safe.

But in the first half of 2005, a "native" mad cow appeared in the United States. This animal was born and raised on a ranch in Texas. The steer was discovered at a pet food processing facility in Waco. Research showed that the sick animal was a twelve-year-old "downer" animal shipped in for slaughter. Downers are sick or aged animals that can't stand or walk and must be put down. These animals are not automatically excluded from animal feed, so often they end up in pet food or livestock feed (not for cattle since 1997, at least not legally) or other products. This mad cow with native BSE may have contracted BSE before the ban on animal products in cattle feed.

This mad American cow caused a considerable stir, briefly.

But on July 11, 2005, Texas authorities, convinced that all problems had been identified and solved, lifted all quarantines and restrictions from the ranch in Texas where the mad cow was raised. The diseased animal was first identified around June 24. Apparently less than three weeks was adequate for completion of the investigation.

In the end, it seemed the only things that we clearly learned were that we still have no way of following the history of individual animals so that we can identify the sources of their feed, and that parts of sick animals are still being fed to pigs—and, perhaps, pets—which is the reason that this downer cow was at the processing plant to begin with. In fact, this mad cow had already passed one inspection and was on its way to becoming pig food before someone noticed that it could no longer stand on its own.

On July 14, 2005, a federal appeals court ruled that the two-year-old ban against importation of Canadian cattle had to be lifted. As recently as March of 2005, a Montana judge ruled that the USDA could not lift the ban on Canadian cattle. At that time, he said that lifting the ban "subjects the entire U.S. beef industry to potentially catastrophic damages" and "presents a genuine risk of death for U.S. consumers."

Regardless, according to David Kravets of the Associated Press, the decision to lift the ban on importation of Canadian cattle came a day after the U.S. Justice Department had urged the appeals court in Seattle to reopen our northern border to Canadian beef imports.

The American Cattlemen's Beef Association was against lifting the ban. Their profits had risen considerably since 2003. The American Meat Institute (AMI) was for lifting the ban. The AMI represents feedlots and meat processors. Since the ban on Canadian beef went into effect, feedlots hadn't been full, and meat processors weren't getting all the meat they could package and sell. The Canadian Cattlemen's Association was also for lifting the ban. The ban had already cost them more than $5.6 billion. A lot of money was changing hands.

Twenty-four hours after hearing from the Justice Department, everyone on the appeals court seemed to agree that the judge who had earlier refused to lift the ban should have paid more attention to the USDA, which had supported the lifting of the ban for some time now. After all, it is the USDA's job to protect us.

The Japanese, though, seem to feel the USDA is not doing an adequate job. According to a July 16, 2005, Associated Press article, a study done by the Japanese Ministry of Agriculture showed that nine out of twenty cases of mad cow disease identified in Japan would have been sent to market if they had occurred in the United States. Japan tests all cattle for BSE before slaughter. The United States tests only those animals that are obviously ill.

And in January of 2006, the Japanese threatened to again ban the importation of U.S. beef because of suspected contamination with nerve tissue.

Importation and feed have always been the problems. There is some evidence that BSE may arise spontaneously in a few cattle. But most all of the documented cases, imported or native, involve feed that contained animal parts. So cattle feed remains the major concern. Because of that, since 1997, feeding cattle with sheep or cattle parts has been against the law in the United States. And because those laws may be all that stand between us and mad cow disease and vCJD, full compliance by ranchers, rendering plants, and slaughterhouses remains a major concern.

During all of the time that it has been illegal to feed cows pieces of other animals, speeding has also been against the law in the United States. But sometimes people are in a hurry. And in case you didn't notice, it appears that it is still okay with the USDA if ranchers feed dead animals, including cows, to pigs.

14 ⦂⦂⦂

An Infectious Holocaust:
HIV

"Excuse me, Mr. Shilts," an early morning caller asked of Randy Shilts on a talk radio program, "I asked are you absolutely sure, if you can categorically state that you definitely can*not* get AIDS from a mosquito."

"Of course you can get AIDS from a mosquito," Shilts replied. And then he paused to let that sink in.

"If you have unprotected anal intercourse with an infected mosquito, you'll get AIDS. Anything short of that and you won't."

It's nearly funny, isn't it—the vastness of our ignorance.

But acquired immunodeficiency syndrome, or AIDS, has killed nearly thirty million of us, infects about forty million of the living, and will kill another three million or so of us this year.

The opening anecdote comes from Rand Shilts's powerful essay, "Talking AIDS to Death," which originally appeared in *Esquire* magazine in 1989. Shilts was the author of *And the Band Played On*, a chronicle of the first ten years of the AIDS epidemic, and a reporter for the *San Francisco Chronicle*. "Talking AIDS to Death" describes many of

Shilts's strange encounters with curious, frightened, and poorly informed people.

It is astonishing, the fog and the fear that surround AIDS. And at times it seems almost amusing, especially when seen through the mischievous eyes of Randy Shilts and through the lens of fifteen years past.

But fifteen years after this essay was written, our ignorance is still monumental, and it's still killing us.

How It Began

Probably, it all started with a chimpanzee, technically *Pan troglodytes troglodytes*—like the chimps we've seen in zoos and on television—infected with the simian immunodeficiency virus, a close relative of the virus that causes AIDS. And it appears chimps have passed this virus to humans on more than one occasion—perhaps as many as three times—beginning at least as long ago as 1959, when Randy Shilts was just seven years old.

For centuries, the peoples of West Equatorial Africa have hunted chimpanzees for food.

A heavy wedge of jungle underlayed with the greenish light of death. A group of young men have treed a chimp. The animal bleeds from wounds to its head and thigh. The men shout and menace the chimp. The air is heavy with water and fear. Suddenly, in a last effort to escape, the chimp jumps at one of the men and bites. The wound is deep. Blood flows freely from both chimp and human. A friend moves in quickly with a spear and ends the animal's life. But it is too late. A virus has slipped between the chimp and the man, and many lives are about to change.

That was one beginning. But chimps appear to be the source of only one type of virus, the human immunodeficiency virus type 1, or HIV-1. Though we now know that HIV-1 has several groups and many subtypes, all of them appear to have come to humans originally from chimps. But there is another type of HIV, called HIV-2. This virus appears to have come to humans—on at least four separate occasions—from a monkey, the sooty mangabey (*Cercocebus atys*).

African peoples also hunted these monkeys for food. Another simian immunodeficiency virus infected sooty mangabeys, and that virus was ready to make the leap from monkeys to humans. Today, HIV-1 is, by far, the most common type of HIV infection. HIV-2 infections are common only in certain areas of West Africa—Senegal, Ivory Coast, Cape Verde, Gambia, Guinea-Bissau, Liberia, Ghana, and Nigeria. And HIV-2 has spread significantly only to countries with ties to West African countries—France, Portugal, Angola, and Mozambique. HIV-2 has spread much more slowly than HIV-1 for several reasons. HIV-2 does not move as easily between people as HIV-1 does. Also, severe immunodeficiency develops much later in people infected with HIV-2. In people infected with HIV-2, the numbers of virus that appear in the blood are rarely as high as those seen in HIV-1 infections. This makes it much less likely that a pregnant woman or an intravenous drug user will transmit the virus.

Both HIV-1 and HIV-2 began as accidents. A simple consequence of hunger and the collision of men, monkeys, and apes. Human hunger has driven the epidemic ever since.

After 1959, the history of this epidemic gets pretty cloudy for the next ten years or so. Clearly, HIV infected some people before 1970. But it seems that through much of the '60s and '70s, the virus just kicked around without stirring up a lot of attention. In the late '70s, though, things got going, and by the beginning of the 1980s, HIV had infected people in North America, South America, Europe, Africa, and Australia. Only then did people begin to take notice.

It took all those years for HIV/AIDS to become a recognized pandemic for several reasons. First, the early symptoms of HIV are relatively nondescript. It can take as many as ten or more years before the really ugly symptoms of HIV infection become apparent. And when AIDS first did show its face, no one knew what it was. Silence was the only sound that accompanied the early spread of this killer around the globe.

But there were warnings. Kaposi's sarcoma is a rare, relatively benign tumor caused by a herpes virus. But by March of 1981, the tumor began to pop up with a greater frequency than ever before, and now this sarcoma was appearing in a more malignant form. Eight cases appeared in New York City alone, all in gay men. At about the same time, homosexual men in California and New York began turning up with a rare fungal infection in their lungs. The infectious agent was *Pneumocystis carinii*, and the disease these men had was *Pneumocystis carinii* pneumonia (PCP).

This fungus infects the lungs of most human beings beginning at about three years of age. Usually, though, the infection does little or no damage. We once thought that *P. carinii* was a protozoan parasite, but DNA analyses have shown that the organism is clearly a fungus. So scientists renamed the organism *Pneumocystis jiroveci*, but we still call the disease PCP. Normally, our immune systems keep *Pneumocystis* under control. But now, *Pneumocystis*-infected men were showing up by the dozens in clinics on both coasts. Something was happening out there, something terrible.

In July of 1981, the Centers for Disease Control took notice of this sudden flare-up of PCP in the United States, and it sounded the alarm.

Though AIDS had been around for decades, it had taken this long for the right people to notice. But even after the alarm bell rang, no one could figure out what they were dealing with. Immunodeficiencies certainly, but what was causing them?

Maybe it was a cytomegalovirus that caused this disease. Cytomegaloviruses are a type of herpes virus that infect somewhere between 40 and 80 percent of adult men and women in the United States, usually with little effect. But we knew that some herpes viruses could be transmitted sexually, and this disease did seem to favor homosexual men. Some herpes viruses also caused immune suppression. That might fit, too. Or maybe it was amyl nitrate, the stuff found in poppers—a prescription drug popular in some gay communities. That could explain why only gay men were getting the disease. Or maybe it was immune "overload" somehow caused by anal intercourse. A lot of possibilities, but mostly these were just random shots into the frightening void. One thing was certain: Whatever was killing these people, it began by shredding their immune systems.

Perhaps to put the majority of us at ease, in July 1981, *The New York Times* reported that Dr. James Curran of the Centers for Disease Control said "there was no apparent danger to nonhomosexuals from contagion . . . no cases have been reported to date outside the homosexual community or in women."

That statement, it turned out, was just another shot in the dark, a horribly misleading one. Within five months, it was clear that this disease was attacking more than just homosexuals. In hospitals in New York and California, intravenous drug users started showing up with Kaposi's sarcoma and PCP. Then heterosexual women who had never injected themselves with anything turned up with fungal infections and sarcomas. Then the babies born to those women developed the same symptoms. And, finally, people who had received blood transfusions started getting sick.

By July of 1982, the CDC had received 452 reports of this new disease from twenty-three different states. And in August, the disease finally had a name—acquired immunodeficiency syndrome, or AIDS. Immunodeficiency diseases arise when human immune systems fail to do their jobs properly. In people with immune deficiencies, immune systems respond to life-threatening infections too slowly, or too weakly, or in the wrong way, or not at all.

Immune deficiency diseases (also called immunodeficiency diseases) come in two types—congenital and acquired. People like David Vetter, the boy in the bubble, have congenital immunodeficiencies, deficiencies that they are born with. Or, somewhere along the way, a person can "acquire" an immune deficiency. Malnourishment, drugs (like steroid hormones), stress, or infection all can cause immunodeficiencies. Early on, the only thing that was clear to everyone about this disease was that infected people weren't dying from whatever started the disease. Instead they were dying from secondary infections that were trivial in people with functional immune systems.

Acquired immunodeficiency syndrome—a starburst of opportunistic diseases—was destroying men and women in nearly every country in the world.

That was the beginning of the end of sexual intercourse as we had known it—a simple pleasure. Sex, abruptly, became dangerous, even lethal.

HIV

In May 1983, Luc Montagnier and coworkers at the Pasteur Institute in Paris announced they had isolated a virus they believed to be the cause of AIDS. Nobody seemed much interested at the time, but the Paris researchers sent the virus to the CDC in the United States for further analysis.

As research on the virus progressed, interest increased rapidly. By April of the following year, the scientists at the CDC were convinced. On April 22, 1984, the Centers for Disease Control announced that researchers there believed they had identified the cause of AIDS. The scientists were speaking of the virus sent to them from Paris.

One day later, U.S. Health and Human Services Secretary Margaret Heckler announced that Dr. Robert Gallo, working at the National Cancer Institute in Bethesda, Maryland, had isolated the virus that caused AIDS. The announcement from the CDC had forced Robert Gallo's hand.

"We hope to have a vaccine ready for testing in about two years . . . yet another terrible disease is about to yield to patience, persistence and outright genius," Secretary Heckler said more than twenty-two years ago.

In her enthusiasm, she had overlooked the impact of ignorance and hubris and how the two would nurture this pandemic in the coming decades.

A battle soon followed between researchers at the National Cancer Institute in the United States and the Pasteur Institute in France. The quarrel centered on who had found the virus first and who second, over who had patent rights and who didn't. That battle would last for years. People were dying, certainly, but a lot of fame and money were at stake here.

Once it was clear that an infectious agent caused AIDS, investigations began to focus on how the virus turned healthy human beings into infection- and cancer-riddled fragments.

We still don't know all of the answers to that question, but we do know a lot more. HIV, the virus that causes AIDS, is a retrovirus. Retroviruses reproduce in a somewhat unusual way. The virus's genetic information is carried by ribonucleic acid (RNA), not the deoxyribonucleic acid (DNA) found in our cells.

When retroviruses infect human beings, the viruses first inject their RNA into our cells. Then the viruses make DNA from their RNA. That step we call reverse transcription, because people and most other living things normally make RNA from DNA. And reverse transcription is the reason we call these viruses retroviruses. The newly made DNA then inserts itself into one or more of the host cell's chromosomes. There the viral DNA can sit quietly for a decade or more before something awakens it from its slumber.

Something—another infection, perhaps—activates the cell that is harboring the viral DNA (ticking all these years like a persistent time bomb), and the cell begins to divide and make new DNA from the genes in its chromosomes. As the cell begins to manufacture human DNA, though, it also begins to make copies of the viral DNA that has been hiding in that human DNA for so long. The new viral DNA then makes RNA. Some of this RNA directs the construction of proteins, which will make new viruses. Initially, the viral proteins appear as one long protein. Then enzymes called proteases cut the long protein into shorter building blocks for assembly of the virus.

HIV infects several types of cells, but most important, HIV infects a group of white blood cells called lymphocytes, particularly T-helper lymphocytes (Th cells). The immune system relies on Th cells to do all the things immune systems do—like keep us alive on a day-to-day basis.

Billions of times every day, bacteria, viruses, fungi, and parasites come after us to use us for food and to make more bacteria, viruses, fungi, and parasites. Even though we're vastly outnumbered in this confrontation, every day most of us survive.

The only thing that stands between any of us and the rest of this hungry world is an immune system. Individuals exist in this world only so long as they can defend themselves against all the other individuals, especially the microscopic ones.

And nowhere is the power of the immune system made more visible than in people with AIDS.

HIV first attacks the Th cells and our immune systems. Oddly, as this begins, not much changes. There is a drop in the number of Th cells and the virus appears in large numbers in the blood. There may be a brief flulike syndrome, but not usually severe. Then immune defenses kick in and we begin to make antibodies (serum proteins) and cytotoxic T lymphocytes (a type of white blood cell that is especially good at destroying virus-infected cells). The antibodies circulate in the blood, and the appearance of these antibodies—anti-HIV antibodies—is the first indication that a person has been infected. Once the antibodies appear, the person is HIV-positive—that is, clearly infected with the virus.

Usually then, everything reverts to normal, or nearly normal. The number of T-helper cells returns to near normal, and the virus drops to very low, sometimes undetectable levels in the blood. A person may live ten or twenty or even more years like this, especially with modern treatments for HIV infection. But all the while, the virus is spreading from Th cell to Th cell and eating away at the immune system. Eventually, a day comes when the immune system moves to strike at some infection and finds that there just aren't enough Th cells to do the job. The immune system begins to fail, and the individual begins to disappear.

Oral thrush—a fungal infection of the mouth and tongue and gums—often is a first sign of AIDS. Many of us have the yeast (a type of fungus) that cause thrush, *Candida albicans*, in our mouths and our digestive systems all the time. But we never notice because our immune systems keep the yeast from overgrowing and doing damage. When immune systems begin to fail, though, the yeast can overgrow explosively and form a thick, white crust across mouths, gums, and tongues of the infected. The underlying tissues begin to ulcerate as the yeast chews into them, and eating, swallowing, and simply living become very painful.

Then other chinks in the armor appear. Kaposi's sarcomas—virus-caused tumors—begin to blossom in brown flowers across people's backs and chests and thighs. *Pneumocystis jiroveci*, another fungus, starts growing inside of people's lungs (PCP). Herpes viruses split lips and anuses. Lymphomas—more virus-caused tumors—spread like ink clouds inside of brains and minds.

In short, as the immune system shuts down, the individual begins to disappear, and in his or her place there rises a community of living things, rich with infection. In the end, the man or the woman evaporates altogether, and where she or he just stood, there is only the fierceness of life itself, the microscopic, the ancient, the frenzy.

Because HIV is so fond of lymphocytes, a good way to transmit HIV is to transfer lymphocytes. And since lymphocytes live in greatest numbers in blood, the very best way to transmit HIV is to transfer blood from an infected person into the blood of an uninfected person.

Blood transfusions move lots of lymphocytes from one person's blood to another's—a very efficient means for transmitting HIV infections. Fortunately, that rarely happens anymore because of all the ways we now screen blood for infection before transfusion. Sharing needles contaminated with infected blood is the next best way to transfer HIV from person to person. Intravenous drug users who share needles also share blood. When that blood contains infected lymphocytes, those drug users share HIV.

Direct transfer of infected blood is *the* most efficient way to transmit HIV, but it is not the most common way. Most commonly we share HIV with one another during sexual intercourse. And among the various forms of sexual intercourse, anal intercourse is the most effective.

How do we get infected during intercourse? There are significant numbers of lymphocytes in human semen (the nonsperm part of the male's ejaculate), and in an infected person, those lymphocytes can carry HIV. Also, HIV appears in urethral and prostate secretions in infected men. So one way or another, semen can be loaded with virus. That provides the means for getting HIV from an infected person into an uninfected person. But that solves only half of the problem—the virus part. How do we get the virus into the uninfected person's blood?

Not always, but often, during sexual intercourse, the mucosa—the epithelial cells lining the vagina or the rectum—can be torn and bleed. This is especially true of the mucosa of the rectum, since this tissue evolved under different types of selective pressures than those that affected the vaginal mucosa. Because of that, rectal mucosa does not withstand the stress of penetration and intercourse as well as vaginal mucosa. So anal mucosa is more likely to bleed during intercourse. The bleeding may be slight and unnoticed, but as we know from the effects of shared needles, it does not take much blood to move enough virus.

Rectal mucosa is also more abrasive than vaginal mucosa, and abrasion can cause scrapes and scratches to the penis, and those can also bleed. That is why anal intercourse is more likely than vaginal intercourse to result in infective viral transmission and why, during anal intercourse, both partners are at risk for infection. But in general, the "receptive" partner is at greater risk than the "insertive" partner.

Of course, the likelihood of transmission of HIV also depends on the amount of virus circulating in the blood of the infected person and how much of that virus appears in various secretions—the more virus, the greater the likelihood of transmitting the infection. It's a complex disease, and predicting the likelihood of infection under any given set of circumstances is very difficult.

For example, in spite of its efficiency, anal intercourse accounts for only about one-fourth of the cases of sexually transmitted HIV infections in the world. And in spite of its relative inefficiency, vaginal intercourse underlies fully three-quarters of the instances of sexual transmission of HIV worldwide.

The Present

More than three decades after it began, the HIV pandemic still rages. And disappointingly, more than three decades after it began, ignorance remains one of our greatest enemies in the battle against AIDS.

A recent study in rural Uttar Pradesh, India, showed that 71 percent of married women (all wed before puberty) knew nothing about sexual intercourse when they married, and 80 percent did not know how a woman became pregnant. Uttar Pradesh is not some backwater. It is a thriving state with much international tourism.

And in Cameroon, Lesotho, Mali, Senegal, and Vietnam, more than two-thirds of young women did not know three ways to decrease the likelihood of becoming infected with HIV. While in Moldova, the Ukraine, and Uzbekistan, over three-fourths of the women asked could not list three methods to prevent AIDS—abstinence, condoms, and an exclusive relationship with a faithful partner.

These facts appeared in a 2004 report by the United Nations and the World Health Organization. In 2004, approximately 40 million people were living with HIV, about 3.1 million people died from AIDS, and about 5 million people became infected with HIV.

In Zambia that same year, only 11 percent of women interviewed thought they had the right to ask their husbands to use a condom—even if they knew their husbands were unfaithful and infected with HIV.

Nor does it appear that issues of bias and lack of solid information are limited to the uneducated peoples of the underdeveloped parts of the world. Writing for The New Yorker in 2006, Michael Specter pointed out that current U.S. "government policy requires that one-third of the HIV-prevention spending go to 'abstinence until marriage,' programs." This is a policy many consider highly questionable.

And, according to Specter, some of the most trusted sources of information in the United States have been manipulated for political purposes. "Several years ago the Centers for Disease Control removed a fact sheet about condoms from its Web site; the sheet disappeared for more than a year, and when it was replaced, instructions on how to use condoms had been supplanted by a message denigrating them. The CDC also removed a summary of studies that showed that there was no increase in sexual activity among teenagers who had been taught about condoms."

Thus, at least some information provided by perhaps the most trusted health and disease Web site in the world now reflects something other than what most scientists consider to be sound scientific and medical advice.

And through their political influence, some groups lead an ongoing effort to put their religious beliefs ahead of well-researched scientific policy. For example, very recently scientists have produced two effective vaccines against human papilloma virus (HPV). HPV causes genital warts and, in many women, appears to be the major factor in the development of cervical cancer. These vaccines represent the first time human beings have produced agents that can prevent cancer. Most of the scientific community is ecstatic and its leaders have proposed that all children be vaccinated against HPV. Some religious conservatives aren't so sure.

According to Specter, religious conservatives not only believe "that the mass use of an HPV vaccine or the use of emergency contraception will encourage adolescents to engage in unacceptable sexual behavior; some have even stated that they would feel similarly about an HIV vaccine, if one became available. 'We would have to look very closely at that,' Reginald Finger, an evangelical Christian and former medical adviser to the conservative political organization Focus on the Family said. 'With any vaccine for HIV, disinhibition'—a medical term for the absence of fear—'would certainly be a factor, and it is something we will have to pay attention to with a great deal of care.'" Finger is referring to a disease that has already infected nearly forty million people and killed almost thirty million.

It is tempting to dismiss Finger's lack of insight and concern over forty million people as simply the uninformed and insensitive opinions of one man. But Finger currently serves on the Immunization Committee for the Centers for Disease Control. That committee regularly makes recommendations to the CDC and the U.S. government concerning vaccine policy.

Furthermore, every year, Brazil spends more than $400 million on AIDS control programs, including a program to distribute free condoms to prostitutes. Until recently, $40 million of Brazil's anti-AIDS funds came from the United States. However, in late July of 2005, Brazil decided it would no longer accept funds from the United States. The United States, it seems, had decided to demand that "all foreign recipients of AIDS assistance must explicitly condemn prostitution."

Pedro Chequer, director of the Brazilian government's AIDS program, said, "Our feeling was that the manner in which the USAID [United States Agency for International Development] funds were consigned would bring harm to our program from the point of view of scientific credibility, its ethical values and its social commitment. We must remain faithful to the established principle of the scientific method and not allow theological beliefs and dogma to interfere."

The New York Times summed it up: "Brazil, which spends more than $400 million annually on what is regarded as the most successful AIDS program in the developing world, is taking a pragmatic approach in combating the global epidemic, the experts say, while the United States increasingly is not."

Mark Dybul, deputy coordinator and chief medical officer for the White House's global AIDS initiative, confirmed that U.S. aid "does require an acknowledgment that prostitution is not a good thing and to be opposed to it."

Brazil has legalized prostitution, and its government believes it would be hypocritical and unethical to brand its prostitutes as evil. For the Brazilians, a more practical and less theological approach is more sensible.

We are battling not only ignorance but social inequality as well.

The highest rates of infection in women occur in countries where marriage is common and most relationships are heterosexual. Men in these relationships often seek out other sexual partners, including prostitutes. Frequently, the sexually active prostitutes have HIV infections. Then, when the husbands return, they bring more than just their ardor back to wives, who feel it is improper to ask their husbands to use condoms. And the cycle completes itself.

Unlike malaria, short of unprotected anal intercourse with insects, we don't get HIV from mosquitoes. That is because female mosquitoes don't inject blood when they bite; they inject saliva. And so far, HIV hasn't figured out a way to colonize mosquitoes' salivary glands.

Even if mosquitoes could transmit AIDS, those insects might well pose less of a threat to human beings than do our own ignorance and the social marginalization of women.

Within the United States itself, a nation with broad education programs and ready availability of condoms and sterile needles, the annual number of new HIV infections has only stabilized—about thirty-two thousand new cases per year. And since people are living longer with HIV, the number of HIV-positive people in the United States continues to rise—currently it is around two million.

After decades of publicity and education, after years of witnessing the carnage this virus has wreaked, U.S. citizens continue to infect one another at a constant rate and continue to die from AIDS.

As depressing as that is, the general state of affairs is even worse. Between 2002 and 2004, the number of people infected with HIV rose in every region of the globe. In East Asia, the number of infected people rose by nearly 50 percent in those two years. It appears that this rise was mostly due to China's HIV epidemic, which has surged. Because of major epidemics in Russia and the Ukraine, in 2004, 40 percent more people in Central Asia and Eastern Europe are living with HIV than in 2002. But worst of all is Africa, especially sub-Saharan Africa: Nearly two-thirds of all cases of HIV infection occur there. And of all the women in the world infected with HIV, more than three-quarters are African. In Africa, the AIDS epidemic has orphaned more than twelve million children under age eighteen—nearly an entire generation of African children without mothers or fathers.

The world is exploding with HIV, and Africa is at ground zero.

In a few regions of the world, the rates of HIV infection have stabilized, but nowhere have the trends reversed themselves. Forty million infected, nearly thirty million dead. Three million died in 2005, five million to die next year. Five million infected this year, seven million next. There is no light at the end of this tunnel.

The Future

An obvious solution to the AIDS pandemic would be an effective and cheap vaccine. Look at what we accomplished with smallpox, measles, mumps, and polio.

But HIV is a different sort of virus. Some thirty years after the beginning of the pandemic, there is still no vaccine against HIV. And it isn't clear that there ever will be one.

As Andrew McMichael pointed out, writing in *Nature Medicine*, with all previous vaccines for other viruses, people who had survived an infection with the virus were clearly immune to later infections with the same virus. But that is not the case for HIV, "because no one is known to have recovered from, and completely cleared, [an] acute infection." Maybe it isn't possible for human beings to develop immunity that can completely eliminate the virus from our systems and protect us from future infections. Maybe the virus is just too well adapted, too fast for our immune systems.

In Nairobi, there is a large group of prostitutes who, despite repeated HIV exposure, have not developed AIDS. Also, partners of some HIV-positive men and women have not developed AIDS. It is not entirely clear that either of these populations is protected by their own immunity or just resistant to infection. But both groups provide cause for hope among those working on a vaccine against HIV. People trying to produce a vaccine, like the infected, are up against a very devious virus.

HIV, like other retroviruses, has a most unusual adaptation that helps protect it from immune attack. When our cells make new DNA from old DNA, they use an enzyme called a polymerase (because it links individual bases into long-chain, or polymeric, nucleic acids). Since any errors that occur during DNA replication (or polymerization) could cause terrible mutations in our genes, our polymerases have what is called an editing function—a means for constantly rechecking the copy against the original and correcting any errors that occur as the new DNA is made.

A retrovirus first makes DNA from RNA. So every retrovirus has an unusual polymerase called reverse transcriptase. This enzyme polymerizes DNA using RNA as a template. Reverse transcriptase has no editing function. Many errors go uncorrected as HIV makes DNA from RNA. That changes the virus's genes. Sometimes, probably most of the time, that's bad for the virus because it destroys some essential viral function. But not always.

Like buying a million or ten million lottery tickets, most of them will be worthless. Among those millions, though, there will be one or two tickets that are worth a lot of money.

Inside an infected person, HIV is replicating itself billions of times inside millions of cells. So even if it is a very rare mutation that proves beneficial to the virus, it is a certainty that that mutation will happen.

Maybe only one in one thousand or one in ten thousand viruses is functional. But sooner or later, a virus appears that is a little different in a very important way. This new virus still does all the things that the old HIV did, but it no longer looks the same to the immune system. To the immune system, this virus looks like a brand-new virus, something never seen before.

When HIV first infects a person, his or her immune system begins to make antibodies and T cells against the virus. These are the weapons of the immune system. Antibodies find threats floating around in the blood or lymph (the fluid in between our cells outside of the blood). And T cells find infected cells and destroy them.

Early on, the T cells, especially, may wipe out a lot of the virus. But before the immune system can complete its job, some of the viruses have changed. In essence, these viruses have restructured their genes and pieced together a new mask. The new mask disguises them so thoroughly that the immune system thinks this is an entirely different virus. So the immune system starts over with an attack on the "new" virus. But by the time the immune system is ready to mount this second attack, the virus has donned another mask. And so on. After a short while, the HIV inside a person with an established infection is not the same virus that originally infected that man or that woman. Also, throughout the course of the disease, HIV is throwing off proteins that act like decoys to the immune system and trick the immune system into wasting its efforts on responses that have no effect on the intact virus.

Add to that the fact that as many T cells respond to the infection, they also begin to produce more HIV from within their own chromosomes where it has remained hidden all these years.

The immune system soon exhausts itself. And, of course, all the while, HIV is killing infected T cells. Against such an onslaught, the immune system doesn't stand much of a chance.

That's also what makes it so difficult to manufacture an effective vaccine—HIV is a moving target, a master of disguises. Somehow, an effective vaccine would have to find a way to keep up with all of these changes, which seems pretty unlikely. The only other way is for us to find some chink in the virus's armor, a molecule that HIV cannot change and still be HIV. Then, if you vaccinated a person, the immunity that developed would still prevent infection by an altered virus acquired from someone else a year or more later. So far, there are no candidates for such a vaccine, no important molecules that HIV cannot change and still infect and reproduce as it always has. Because of that, the vaccine route appears thorny at best.

And even a vaccine would not clearly put an end to all of this. We have a highly effective vaccine against measles. Still, every year, nearly three-quarters of a million children in developing nations die from measles. Our very best technologies pale in comparison to the force of poverty.

Therapy of HIV—that is, treatment of people with active infections to slow the progress of their disease—has made some significant strides.

The current state-of-the-art treatment is highly active antiretroviral therapy, or HAART.

Ideally, therapy of HIV infections should help to destroy viruses without destroying healthy human cells. HIV does two things our cells don't do. First, it uses reverse transcriptase to make DNA from RNA, and second, HIV uses proteases to chop its one large protein into the shorter pieces that it needs to build new viruses. Both enzymes, reverse transcriptase and HIV's protease, are most active in cells producing the virus. HAART uses both of these unique viral events to slow the progression of AIDS. HAART drugs include a mixture of three or more anti-HIV drugs. Often these include two different types of reverse transcriptase inhibitors (such as AZT) and a protease inhibitor. Inside an infected person, these drugs interfere with HIV's ability to make DNA from RNA and to cut out the protein building blocks it needs to make more virus. That helps to slow the infection and at least partially restore the immune system. The time from infection to the development of AIDS increased 14 percent (among the least ill) to 34 percent among the sickest women and 40 to 100 percent in men. Though the reasons behind these sexual differences aren't entirely clear, it is obvious that some people have benefited a great deal more than others from this therapy. But HAART is expensive, costing $1,000 a month or more. And expensive or not, it often simply isn't available to some of the most affected populations, like the rural poor of Africa.

That leaves prevention as the single viable option—prevention through education. People need to understand the danger of sharing needles between intravenous drug users, the insanity of risky sexual behavior, the vital importance of condoms, the need for fidelity, and the importance of the empowerment of women. Ignorance is a great enemy.

But even in the clinics in some parts of Africa, the medical staff reuses syringes and needles. They know better, but they have no choice. There is no money to buy more needles. Ignorance is a powerful enemy; poverty is an even stronger one.

Add to that the deeply rooted and essential human sexual drive, and you have a combination for continuing disaster.

In Africa, even if a vaccine was developed tomorrow (which it won't be) and made available to all the poor of that nation (which it won't be), HIV and AIDS would (will) still kill tens of millions more, because of all of those who are already infected. No matter what we do, there will be nations of orphans.

When Randy Shilts wrote "Talking AIDS to Death," he already knew that he was infected with HIV. He learned of that infection in 1987 on the day he wrote the last page of *And the Band Played On*. For a few years, drugs held his disease in check. But in 1992, on the eve of his forty-first birthday, he developed AIDS-related pneumonia. Four months later, his left lung collapsed. In 1993, he developed Kaposi's sarcoma, and a year after that, in 1994, Randy Shilts died from his HIV infection. Just like Arthur Ashe, Isaac Asimov, John Holmes, Rock Hudson, Roy Cohn, Liberace, Robert Mapplethorpe, Rudolf Nureyev, Anthony Perkins, and Alan Wiggins—a man who ran like the wind for the San Diego Padres—died from their infections. Just like another eight thousand people will die from their infections tomorrow, and the next day, and the next. . . .

15

The Slate Wiper: Influenza

I had a little bird
Its name was Enza.
I opened up the window
And in flew Enza.

My great-grandfather came to Kansas with the railroad—not in the dining car, but pushing a handcart, driving spikes, blasting roadbed, and laying rails. Along the way, my father's grandmother bore five children, four boys and one girl. One of those boys was Charles Gerald, my grandfather, the one who gave his leg back to the railroad in return for his passage to Kansas. Another child was Cornelius, or Con. As a boy, my father's job on paydays was to find his uncle Con and pull him out of whatever gin joint he'd set up shop in before Con could spend all of his pay. My father did his best, but it wasn't good enough. Con managed to kill himself with alcohol almost as soon as he was able.

Great-uncle Frank I don't know much about. Maybe that's because

Frank didn't distinguish himself with drink, dementia, or dismember-
ment as the others had. Or it might have been something darker. Regardless, Frank's name has disappeared from the Callahan family
annals. Margaret, the lone girl, was a wonderful woman who fed me
and my brother and sisters too many oatmeal cookies and some won-
drous lies. Blue-eyed, with a smile like a fistful of diamonds. Alzheimer's
disease—though no one called it that back then—took Great-Aunt Margaret's mind, a piece at a time, until there was nothing left worth knowing. The family kept her in back rooms of various houses until her heart
finally gave out.

Craziness, in most of its manifestations, and catastrophe run deeply
in the roots of my family, on both sides.

But even among such an odd assortment of the crippled and the
crazy, one child stood out—my father's uncle Will. William Patrick
Callahan was born stone deaf. Because of poverty and ignorance and a
general lack of interest, Will never learned to speak.

Of them all, Will fascinates me most. Imagine a life without stories.
Most of us start to learn words and stories the instant our ears open—
even while we are still inside our mothers' wombs. Stories create us, and
they create the world for us. How would a shaft of wheat look to someone who knew nothing of seeds, a mother without a childhood, a father
who never touched him? A storm? A gravestone? A world without
words.

I've seen pictures of Will. He was dark-haired and pale. Even in the
old black-and-white photos, I can tell that. His eyes were full of cloud-
less skies. He was thin-armed and bony-kneed. Dressed in overalls and a
faded plaid shirt. I wish for all the world I could have met my great-
uncle Will. And I wish for all the world that Will might have learned to
speak and told someone what he thought about before he had words—
how things looked and felt without "red" or "ripe" or an inkling of
death.

In 1918, when Will was twenty-one years old, he fell ill with the
grippe. My great-grandmother worried herself sick over Will, but it

made no difference. In a week's time, Will was dead and gone without a word. They buried him beneath an early snowstorm on the bleak Kansas plain.

Influenza killed Will. The Spanish flu, as people called it then, though Spain had little, if anything, to do with it. Regardless, it sealed Will's lips and ears forever. And before the Spanish flu was finished, it was responsible for the great influenza pandemic of 1918. A time of true terror. And no one saw any of this coming, even when the signs began to get pretty obvious.

On March 4, 1918, at Fort Riley's Camp Funston in Haskell County, Kansas, the company cook Albert Kitchell reported to the infirmary with a sore throat. His temperature was 103°. At that point, Camp Funston held 60,000 soldiers. Later that day, Corporal Lee Drake and Sergeant Adolph Hurby also showed up at the infirmary complaining of fever and aches. The next day, 40 more men came up ill. By the end of the week, 522 soldiers had come down with the flu. Forty-six of those men died.

Although that was a lot of sick people, the number of dead wasn't extreme, at least for the Army. For the rest, the flu was relatively mild; most of the men recovered in a few days. So the medical staff didn't think too much about it. By the beginning of August, things were pretty much back to normal. About then, the troops from Fort Riley began shipping out for the European theater. They brought their duffels, their rifles, some extra rations, and photos of their sweethearts. They also brought the flu.

France, Germany, Britain, and Spain, especially Spain, soon reported hundreds of cases of influenza, only this flu was worse than the one that hit Camp Funston in March, much worse. And it was worst of all in Spain. Perhaps that's why it came to be called the Spanish flu. Among the soldiers, the flu took an especially terrible toll. In some of the camps, as many as 30 percent of the infected died, mostly from pneumonia— raging inflammations of their lungs. Bacteria went after those who didn't die right away and, in many cases, finished what the flu had begun.

Later, when the soldiers came back home, they brought this newer,

THE SLATE WIPER • 239

more virulent flu with them. In September of 1918, there were 133
cases of flu in Kansas. After the first week of October, there were 1,100
cases. Two weeks later, 12,000. That October at Camp Funston, 14,000
people got the flu, and 861 of them died. The bodies were laid in
hangars like cordwood while they waited for their coffins. By the end of
1918, 12,000 people in Kansas had died from the flu. One of those, of
course, was my deaf-and-dumb great-uncle Will. There were so many
dying, though, that practically no one, save Will's immediate family,
noticed Will's death or the fact that along with Will went all his untold
stories.

All the while, American men continued to ship overseas to fight the
war.

In Brest, France; Boston, Massachusetts; and Freetown, Sierra
Leone, people started dying.

North America, Europe, Asia, Africa, South America, India, and the
South Pacific—relentlessly, the virus circled the globe. Steam shovels
scooped holes in the earth and filled them with dead people.

HIV killed 25 million people in its first twenty-five years. The 1918
influenza killed that many people in twenty-five weeks. A year or so
later, when the flu was done with the world, it had killed 40 million to
60 million people—more people died in that one year than had died
in the four years of the black death (bubonic plague) from 1347 to
1351. In 1918, nearly half of those who died in the war died from influenza. More than 700,000 Americans died that year—more Americans than died in World War I, World War II, Korea, and Vietnam
combined.

And then it was gone. Eighteen months of hell, then nothing. Silence. But in its wake lay tens of millions of bodies, mostly young. Unlike other flus that killed the very young and the very old, the Spanish
flu ripped the heart out of the middle—most of the people who died
were between the ages of fifteen and thirty-four. And unlike other flus,
the Spanish flu killed quickly and efficiently. Infected people died
within weeks, not months. In 1918, life expectancy in the United States
dropped by ten years. Worldwide, nearly 3 percent of those infected

died, whereas other flus killed only about 0.1 percent of infected people. Of course, in some places, things were much worse than that. In some places, 100 percent of those infected died.

In the fall of 1919, a third wave of the flu swept through the United States and the world. But this flu was not nearly as bad. A few died. And then we began to forget.

If the Spanish flu happened again tomorrow, it might kill two hundred million people, maybe three hundred million worldwide — equivalent to the entire population of the United States.

Of course, it won't happen again. After all, that was nearly one hundred years ago. Look at all we've accomplished since then: medical care, flu vaccines, antiviral medicines. Those things will make a difference, a huge difference. Won't they?

Maybe.

The Influenza Virus

We call the worst of the flu viruses influenza A. Like HIV, influenza A has an envelope. And like HIV, influenza A uses RNA as its nucleic acid. But the similarities end there. Influenza A is not a retrovirus, and it doesn't hide out in its host's DNA or use reverse transcriptase. But the polymerase influenza A uses to make more RNA is also error prone and has little editing capability.

That makes it possible for the virus to change its appearance quickly. And beyond its penchant for mistakes during RNA replication, this virus has evolved a whole series of other means for foiling human immune systems. Because of that, influenza A has killed many more human beings than HIV, and most likely it will continue to do so.

Under a microscope, an influenza A virus looks a little like a kidney bean surrounded by one of those knobby rubber balls we give our dogs to chew on. The kidney bean is the viral part; the knobby rubber ball is the envelope part.

Inside the kidney bean are eight strands of RNA—the genetic information needed to make more viruses. The rubber ball is made up of lipids (fats) that the virus took from the host cell it infected. Stuck in the rubber ball are spikes of proteins we call hemagglutinin and neuraminidase. The influenza virus uses these protein spikes to infect our cells. We use these protein spikes to classify the different forms of the influenza virus. We know that flu viruses use at least fifteen types of hemagglutinin molecules and nine types of neuraminidase molecules. So we use these molecules to sort out different forms of influenza A. For example, the Spanish flu of 1918 was an avian flu designated H1N1 — H for hemagglutinin and N for neuraminidase. The 1968 Hong Kong flu pandemic was caused by an H3N2 variant of influenza A. And the avian flu that is causing such a stir right now in China, Southeast Asia, Indonesia, and Russia is an H5N1 variant. There are only four H:N combinations among flu viruses that most commonly infect human beings—H1N1, H1N2, H2N2, and H3N2.

Differences in the hemagglutinin and neuraminidase molecules determine which species of animals the viruses can infect and just how much damage they may cause afterward. The horror of the 1918 flu appears to be most directly attributable to the particular version of the hemagglutinin molecule (H1) it carried, though all the factors still aren't clear.

There have been two major influenza pandemics since 1918—one in 1957 and one in 1968. Because the science was better in 1957 and 1968, we know that both of these pandemics began with a bird virus. That virus found its way into an animal, probably a pig, that was already infected with a human influenza virus. The two viruses then recombined their genes to create a new virus with some of the properties of both of the other viruses—the virulence of the avian virus and the infectivity of the human virus. Influenza pandemics followed.

But in 1918, we knew little or nothing about the nature of flu viruses. So until recently, we could only guess at the origins of the 1918 virus. What most modern scientists guessed was that this virus—like the viruses responsible for the 1957 and 1968 pandemics—began in birds and then got worse when bird and human viruses recombined inside of pigs. That was what we thought until 1997.

In 1997, a group of scientists opened the grave of an Inuit woman who had died in Seward, Alaska, from the flu in 1918. The woman's grave lay in the Arctic permafrost, so everything that had been laid to rest there in 1918 was still there, including the flu. From the corpse, the scientists scraped out bits of lung. From the lung tissues they isolated RNA. From the RNA they made more RNA and injected it into mice. The mice died very quickly, but not before the scientists were able to isolate fully functional 1918 influenza A virus—literally raised from the dead.

When scientists began to pull apart the secrets of the virus, they found that it had not come from some recombination between an avian and a human virus. Instead this virus had jumped directly from birds into humans. And second, these researchers found that the 1918 virus was at least one hundred times more effective at killing mice than any other virus examined. All the reasons for that are still not clear.

Infected persons exhale influenza A virus in tiny droplets when they cough or sneeze. If another person inhales those droplets, the virus infects that person, too. Inside the new host, the virus quickly attacks the respiratory epithelium, the cells that coat the inside of the nose, mouth, throat, and bronchi of the lungs. The virus strips away the mucous layer that protects the epithelial cells and then fastens itself to those cells using its hemagglutinin. The virus fuses its envelope with the host-cell membrane and slides inside the host's cells. There it sets about the task of making new viruses. An uninfected cell devotes all its metabolic activities to keeping itself healthy, energetic, and reproducing. But inside of an influenza-virus-infected cell, the virus co-opts the cellular machinery to make more viruses. And those viruses, once complete, bud through the host cell membrane—taking sizable chunks with them as their envelopes—and go looking for more cells to infect. Because of what the virus is doing, infected cells begin to suffer. As new viruses are made, there are not enough energy and proteins and nucleic acids left over to keep the host cell healthy. Epithelial cells begin to die, and that triggers inflammation. Inflammation causes the blood vessels to swell and fluid to leak out of those blood vessels. Along with those fluids come cells and molecules that push the inflammation even farther. Swelling, itchy eyes, fever, aches, pains, and sore throat all follow.

As the virus bursts into the blood and lymph, it triggers the immune system, and the race is on. If the virus wins, the person may die. If the immune system wins, everything is roses.

Most of this battle begins in the nearest lymph nodes, usually the ones just under our jawbones. These become the swollen "glands" of influenza. Cells of the immune system, called dendritic cells, latch hold of the virus, process it, and hand it over to lymphocytes. The lymphocytes begin to divide, and soon there are millions of them. Some produce antibodies; others go off in search of infected cells. Antibodies are not much use against an established infection, because antibodies cannot get inside of cells. So the outcome of the battle depends on the lymphocytes' ability to find and destroy infected cells.

Viruses multiply, the inflammation in the lungs spreads. Fluid begins to fill the spaces where the vital exchange of oxygen and carbon dioxide normally happens. Breathing becomes more difficult. The immune system strikes back. If the immune system is too slow, the lungs continue to swell and fill with fluid. Eventually breathing becomes labored and then impossible.

If the immune system is faster, the inflammation begins to recede. The lungs empty, and breathing again becomes normal.

The 1918 flu was very fast, fast enough to kill about 3 percent of the people it infected. Both the polymerase the viruses used and the viruses' ability to attach to cells using only their own enzymes appear to be at least partially responsible for their speed and deadliness.

In some people, lymphocytes do their jobs very well, and in others, not so well. So as the flu sweeps through a human population, some people hardly notice, others have runny noses and fevers but end up fine, and others die. Somehow, all of that is predetermined by the nature of the virus and the nature of human immune systems. It is impossible to predict just who will or will not fail. Generally, the immune systems of the elderly and the very young are the weakest. So, often we can predict that influenza will hit these populations hardest. But as we learned with the 1918 flu, even these sorts of predictions may be useless when it comes to flu pandemics.

Normally it takes the lymphocytes of an average human being a week to two weeks to reach full strength. During that time, the flu virus is making more flu virus. So from about the time we experience symptoms, we begin to shed virus. Every time we cough or sneeze or kiss a loved one, we share those viruses. We continue to do that for about the next five to seven days. During that time, we are infectious or contagious. Then, if all goes right, the immune system kicks in and begins to destroy the sources of the virus. We stop shedding virus, and we begin to recover.

And we're fine for a year or two. Then it starts all over again.

Why is that? When we get over the mumps or the measles, we are good for life. We will never again get either of those nasty infections. Why isn't it the same for flu?

If we look at the number of illnesses and death from flu over the years, we find that every three years or so, the numbers spike noticeably. Fortunately, the 1918 spike has never quite repeated itself. But in 1968, there was a major outbreak of flu. It was pretty much over by 1969. By 1972, the flu was back, not as bad, but back. Once more in 1973 and again 1975.

What's going on here? Clearly, by the end of 1969, most of the people still walking around were immune to that year's flu. How come a bunch of people got sick again in 1972? There was absolutely no reason to believe that these people's immune systems had failed or that the immunity they built up in 1969 was gone. So if anything had changed, it must have been the virus.

Influenza A virus has the power to change itself. And it can do that with surprising speed. The change involves two processes called antigenic drift and antigenic shift. Antigens are the parts of the influenza virus that trigger immune responses. So the immune system notices when antigens shift or drift.

Each time influenza A RNA makes more RNA, errors occur in the sequence of the new RNA. That's because the RNA polymerase of the influenza virus isn't very good at what it does, so errors occur while copying the RNA. The polymerase isn't much good at editing those errors out either, so they persist. As a result, the new RNA is often different from the old RNA. Over time, as the changes in the RNA accumulate, the flu virus begins to look different. So different that, after a year or two, human immunity to an old flu is no longer of any use against new flu, and a flu outbreak happens again. How bad that outbreak is depends on the nature of the changes in the virus. We call this process antigenic drift, and it slowly, but inevitably, strips away the face of the old virus and replaces it with something human immune systems have never seen. The old immunity to flu becomes useless, and our bodies have to start defending us all over again. That takes time, and time is exactly what the flu virus needs—time to make more of itself, time to move from person to person—before immunity kicks in and begins to destroy the virus.

Antigenic shift causes change in the virus much more quickly. It is not unusual for more than one type of flu virus to infect an animal or a human at the same time—say, a human and an avian form of the virus all at once inside of a pig or a person—like flu viruses that caused the 1957 and 1968 pandemics. When that happens, the flu virus has the opportunity for dramatic change. As both viruses strip off their coats and begin replicating themselves, it is possible for whole chunks of RNA to move from one flu virus to another. What may emerge is a once-avian virus that can now infect humans. Or, inside of humans or pigs, such a virus might recombine with another human flu virus and suddenly acquire the ability to jump from human to human. This is the great fear about the H5N1 flu that is brewing in Asia. Because once a flu develops the ability to move from human to human, things quickly become ugly. This is antigenic shift, and its power to change the flu virus is nearly unlimited. Again, the face of the virus changes. But this time, not only do those changes perplex human and animal immune systems and buy the virus the critical time it needs, but these changes also rewrite the story of who gets infected.

It is now clear that some avian influenza viruses can jump directly from birds to humans. In fact, this has happened at least ten times in the last ten years. This adaptation may involve antigenic drift. Regardless, infection of humans by avian influenzas is not uncommon. And it some cases, like in 1918, it can result in extremely virulent strains of influenza A. The resurrected 1918 flu virus produced fifty times more virus particles than other strains of influenza A when it was grown in human lung cells, caused mice to lose 13 percent of their body weight within two days of infection, produced thirty-nine thousand times more virus particles than other influenza viruses after four days in mouse lungs, and killed every infected mouse, while another strain of influenza A killed none.

Influenza has been dealing with mammalian immune systems for millennia. Along the way, it has developed a few tricks. Those tricks create rapidly moving targets for immune systems and vaccines.

Each year, the World Health Organization tries to produce a vaccine that will anticipate the coming year's epidemics and protect at-risk populations (usually the very young and very old) from the new year's flu. Each year, that vaccine contains two types of influenza A (an H3N2 and H1N1) and an influenza B virus. Influenza B viruses generally cause much more localized outbreaks of flu and do not change as rapidly as influenza A viruses. It is not an easy job to predict what influenza viruses have up their tiny sleeves. And our hit-and-miss efforts have produced some unexpected results.

Among the vaccinated elderly, the incidence of flu-related deaths seems to be increasing rather than decreasing, as expected. A study done at the National Institutes of Allergy and Infectious Diseases found that flu-related deaths among people over sixty-five years of age have increased significantly since 1980—despite a concurrent rise in the percentage of elderly people vaccinated from 15 to 20 percent in 1980 to 65 percent in 2001. None of these deaths were the apparent result of vaccine-induced influenza, since the most commonly used vaccines contain only inactivated viruses.

Viruses and vaccines don't always do what we ask of them. Nor can we ever guarantee that the virus around the corner is going to act like any virus we have seen before.

The Coming Influenza Pandemic

It isn't a matter of *if*, only *when*. The nature of influenza (antigenic drift and shift and who knows what else) ensures that we will see more pandemics, more worldwide illness and death. The hard part is figuring out when.

But now looks like a pretty ominous time.

In May of 1997, a young boy appeared at a clinic in Hong Kong. He had terrible chest pain and went into respiratory failure. Within a few days he was dead. An autopsy uncovered nothing. Influenza seemed an obvious possibility, but none of the tests for known human influenza viruses came up positive. In August of that year, someone finally thought to test the boy's tissues for other types of influenza. Those tests found a flu virus normally seen only in birds, particularly chickens and ducks. This virus, an H5N1 variant, was first isolated from a bird in South Africa in 1961, but the virus had never been known to infect humans.

At the day school the boy attended, a teacher kept chicks for the children to play with. Some of those chicks had died for no apparent reason. At the time, Hong Kong was experiencing an epidemic outbreak of influenza in its chickens. Clearly, the story of influenza H5N1 was in the process of rewriting itself.

Within a few weeks, medical workers found this H5N1 in another eighteen people. Six of them died. That's a kill rate of over 30 percent. Remember, the kill rate for the 1918 flu—the worst flu in recorded history—was less than 3 percent. Hong Kong closed its borders with China.

The poultry markets of Hong Kong ran red as public health workers slaughtered and burned 1.5 million chickens. Birds don't get influenza like humans get it. First of all, birds get it mostly from the feces of other infected birds. And inside of birds, the virus does something much worse than it normally does in humans. In birds, the H5N1 influenza causes a hemorrhagic fever. Within a very short time after infection, the blood vessels in the trachea and the intestine burst, "turning a pen of healthy birds into a blood-mass of goop and feathers within 24 hours."

Public health officials held their collective breath. Everything calmed down. People stopped dying, and the chickens got healthier. Hong Kong reopened its borders with China.

Most everyone, from the chicken farmers to the scientists, suspected that China had been the source of Hong Kong's H5N1 flu epidemic, but nobody could prove it, and China wasn't speaking about it.

In 2003, two more cases of H5N1 flu appeared in humans in Hong Kong. These two people were members of a family that had recently visited China. One of the two died. A third family member died from a respiratory illness while still in China. But no testing was ever done, so the cause of that third death remains unknown.

In January of 2003, the World Health Organization reported serious outbreaks of H5N1 influenza in poultry in Asia. As more chickens fell victim to this virus, people began getting the disease as well. Between December 30, 2003, and March 17, 2004, Thailand reported twelve cases of H5N1 human influenza, and Vietnam reported twenty-three cases. Twenty-three of the thirty-five infected people died.

Sporadic cases of human infections have continued to appear from Vietnam to Russia to Africa. As of March of 2006, H5N1 influenza A had officially infected 204 people, and 113 had died. There are reports that many more, perhaps as many as 300 deaths from avian flu in China, have gone unreported.

But we still do not know exactly how this influenza is killing people. Autopsies have found the virus in the lower respiratory tract and the gastrointestinal tract. That is in marked contrast to infections with classic human influenza viruses. In humans, traditional human influenza viruses replicate only in the upper respiratory tract and never in the intestine. The effects of the H5N1 virus in the gastrointestinal tract are not clear, since only the lungs from these patients showed signs of marked inflammation. So, it doesn't appear that the virus is infecting and destroying cells in the intestines of people.

Regardless of how this flu is killing people or what is going on behind the scenes in China, one thing is clear: The kill rate for this H5N1 influenza virus in humans is in excess of 50 percent—that's nearly twenty times more deadly than the 1918 flu, the deadliest flu we have ever known.

Epidemic outbreaks of H5N1 influenza in poultry are now occurring in Cambodia, China, Russia, Kazakhstan, Indonesia, Japan, Laos, South Korea, Thailand, Vietnam, central and northern Africa, and parts of Eastern and Western Europe. More than 100 million birds have died in China and Southeast Asia, more than 120,000 in Russia, and more than 9,000 in Kazakhstan.

The sudden spread of this H5N1 virus suggested to many that migratory wild birds were also infected. In late August of 2005, Russia, Mongolia, and Kazakhstan reported H5N1 infections in migratory birds. At the end of April of 2005, China had already reported the H5N1-related deaths of more than six thousand bar-headed geese at Lake Qinghai in north-central China. That event was the largest wild-bird die-off from influenza ever recorded. Perhaps the flu killed all of the birds it infected. Perhaps not.

Birds that survived the flu, though, began migrating over the Himalayas into Myanmar, India, and Pakistan in June of 2005. The same virus found in the geese at Qinghai also proved to be much more lethal in mammals. And clearly, because the virus ends up in feces that end up in lakes and streams, this H5N1 virus will have access to mammals.

If any infected birds survived, they took the virus with them over the mountains. Because of that, Ward Hagemeijer of the conservation organization Wetlands International in Wageningen in the Netherlands told *Nature* magazine, "The next step we expect the virus to take is into

MICROBES THAT WILL CHANGE THE WORLD • 250

Africa, because that is on the main migration route for the birds. The impact in Africa will be dramatically different from the impact in Europe." This is because, according to *Nature*, "the rural communities around the Rift Valley region depend heavily on poultry to survive, and [the people] live in close contact with both domestic and migratory birds." It should soon be apparent just how many migratory birds are still infected. As of spring 2006, infected birds had been found in many parts of Africa.

Because of that and Russia's belated admission that it had a major epizootic of avian influenza including migratory birds in Siberia, it became clear that all of Europe should brace itself for major outbreaks of avian flu as soon as the birds begin moving. Already, the virus has been steadily moving westward through Siberia's Novosibirsk, Tyumen, Omsk, Kurgan, and Altai. As the birds move south for the winter, they will travel through Azerbaijan, Iran, Iraq, Georgia, Ukraine, and several Mediterranean countries. The virus has now reached Europe.

In January of 2006, Turkey reported a major outbreak of avian influenza, and by the middle of that month fourteen people had died. The virus appears to be right on schedule.

More than 150 million birds have now died in Southeast Asia. Human cases of avian flu virus infection have occurred in Vietnam, Thailand, Cambodia, and Indonesia.

The H5N1 flu has also turned up in domestic house cats and captive tigers. It even appears that, inside of tigers, the virus has recombined to generate a form of the avian flu that can be passed from tiger to tiger.

Herons, geese, swans, brown-headed gulls, black-headed gulls, great cormorants, ducks, ostriches, crested hawk eagles (two of which showed up in Belgium), and hawks have all been infected.

At the end of August 2005, H5N1 was found in pigs in China. This is particularly worrisome, since pigs have receptors for both avian and human influenza viruses. That means that in pigs, the potential exists for recombination between avian and human flu viruses—antigenic shift. And the virus produced by such a recombination might have the capacity to move directly from human to human.

Most frightening of all, there have been reports that suggest possible cases of human-to-human transmission of H5N1 influenza A virus. Forty-one of 109 cases identified between January 2004 and July 2005 occurred in families, with 2 to 5 cases occurring in each family. Of course it is possible that that simply reflects the fact that all families were exposed to the same sources of virus. But in three influenza clusters in Vietnam, more than a week elapsed between the first and second illnesses. While there are many explanations for these events, the simplest is that one family member transmitted the H5N1 flu directly to another.

Another remarkable feature of this pandemic is that it appears to have resulted from at least three and possibly four independent epidemics. The first began in August of 2003 in Indonesia and continued at least until October of 2004. The second began in December 2003/January 2004 in Japan and South Korea. The third began in late 2003 in Thailand, Laos, Cambodia, and Vietnam. This epidemic died down after March of 2004 but then flared again in July 2004. And the epidemic in China followed a similar course to that seen with the epidemic in Japan and Korea. Whether or not this Chinese epidemic is truly a separate epidemic remains moot, because the Chinese government has blocked some scientists' efforts to characterize the H5N1 virus involved in the epidemic there.

In the meantime, the virus isn't standing still. Since the pandemic began, both the virulence (how often infection causes disease) and the infectivity (the frequency of infection after exposure) of the virus have varied.

Infectivity of the viruses isolated early during the pandemic was relatively low. More recent isolates have considerably higher infectivity. And while the virus has maintained high virulence for chickens over the ten years of the pandemic, more recent strains of the virus have shown higher virulence for humans. Also, early viral isolates had low virulence for ducks; the latter ones are more pathogenic.

Very recently, scientists have isolated variants of the H5N1 avian flu that are resistant to oseltamivir, better known as Tamiflu. This is obviously of considerable concern, since Tamiflu is one of very few agents that appear to be effective for treatment of H5N1 avian influenza.

Infected birds shed influenza A in their feces. That shed virus can survive in lake water at 8°F for many months and even longer at −15°F. And the virus survives repeated freezing and thawing. That means that the great lakes of northern Siberia are now incubators for H5N1 avian flu and will continue to be for some time to come. Birds who drink that water will be at risk for months more.

The virus also survives for days to weeks in meat from infected pigs, and it is likely that the virus will also survive in poultry meat. Cooking does destroy the virus, but there is still considerable risk from handling contaminated meats before cooking. It is remarkably easy to move a virus from hand to mouth during meal preparation. And there are obvious risks involved with eating raw meats or other animal products such as duck blood.

Viruses wait for no one, especially flu viruses.

The Slate Wiper

Epidemiologists have called influenza the "slate wiper" because of its virulence and its potential for rapid spread across the world. Once the 1918 flu found its way into people and began to move from human to human, the flu spread around the world in just one year—and that was before jet planes, before commuter trains, cruise ships, and automobiles moving across countries at eighty miles per hour.

Still, that flu managed, in a few months, to infect nearly one-half of the world's population and to kill 40 million or so of us. Today, one-half of the world's population is about 3 billion people. Two and one-half percent of 3 billion is 75 million. Fifty percent of 3 billion is 1.5 billion people.

In August of 2005, Anthony Fauci, the director of the National Institutes of Allergy and Infectious Diseases, announced that preliminary data from tests on 450 healthy adults indicated that a new vaccine caused an immune response that appeared strong enough to protect people from the avian H5N1 flu virus. He announced that the government had already acquired two million doses of that vaccine from a French manufacturer and planned to acquire significantly more from American manufacturers as soon as possible. If the vaccine is as good as he hopes, in the case of a true pandemic, two million doses would protect about 1 percent of those who might have become infected and would decrease the death toll on the United States by about one million people. Instead of seventy-five million (50 percent infection rate and 50 percent kill rate), we would lose only seventy-four million of our friends, lovers, sons, and daughters.

Such numbers of dead are beyond our capacity even to imagine. Our minds can envision the death of a few hundred, maybe even conjure the image of a few thousand bodies. But human brains never evolved the means for comprehending the wholesale devastation of our species. We have been so many and so dominant on this planet for only a very short period of time, not nearly long enough—if any amount of time is ever long enough—to acquire the minds we would need to imagine hundreds of millions of people so casually erased by a thing no human eye will ever see.

Like my great-uncle Will, we are standing in the path of a firestorm we can do nothing about, not even imagine. All our lives we have relied on our stories to carry us through—for security in times of danger, for solace in times of war, for direction in times of uncertainty, for peace in times of loneliness. But words and stories have limits. In the face of the virulence and the sheer numbers of the microscopic, our stories crumble.

Confronted with such magnificence, we, like my great-uncle Will, are abandoned by words. In their place, we find—as he did—only a great and silent awe.

Notes :::

1: Infections

9 Premature delivery more frequent in absence of *Lactobacilli*: R. Usui et al., "Vaginal *Lactobacilli* and Preterm Birth," *Journal of Perinatal Medicine* 30, no. 6 (2002): 458–66.

Human milk contains proteins that support the growth of *Bifidobacteria*: C. Liepke et al., "Human Milk Provides Peptides Highly Stimulating the Growth of *Bifidobacteria*," *European Journal of Biochemistry* 269, no. 2 (2002): 712–18.

Sequential bacterial colonization of the human gut: C. F. Favier et al., "Molecular Monitoring of Succession of Bacterial Communities in Human Neonates," *Applied Environmental Microbiology* 68, no. 1 (2002): 219–26.

10 High density of bacteria in the gut: G. J. Tortora, R. F. Berdell, and L. C. Case, *Microbiology: An Introduction* (San Francisco: Pearson Benjamin Cummings, 2004).

Bacterial composition of human feces: F. Guarner and J. R. Malagelada, "Gut Flora in Health and Disease," *Lancet* 361, no. 9356 (2003): 512–19.

Size of bacterial genome found inside humans: L. V. Hooper, T. Midtvedt, and J. I. Gordon, "How Host-Microbial Interactions Shape the Nutrient Environment of the Mammalian Intestine," *Annual Review of Nutrition* 22 (2002): 283–307.

11 Percent of total human DNA derived from bacteria: F. Backhed et al., "Host-Bacterial Mutualism in the Human Intestine," *Science* 307, no. 5717 (2005): 1915–20.

Children born in different hospitals have different gut flora: Guarner and Malagelada, "Gut Flora in Health and Disease."

Differences in normal flora of formula-fed and breast-fed babies: D. Dai and W. A. Walker, "Protective Nutrients and Bacterial Colonization in the Immature Human Gut," *Advances in Pediatrics* 46 (1999): 353–82.

12 Individual differences in the composition of human normal flora: E. G. Zoetendal, A. D. Akkermans, and W. M. De Vos, "Temperature Gradient Gel Electrophoresis Analysis of 16S rRNA from Human Fecal Samples Reveals Stable and Host-Specific Communities of Active Bacteria," *Applied Environmental Microbiology* 64, no. 10 (1998): 3854–59; P. B. Eckburg et al., "Diversity of the Human Intestinal Microbial Flora," *Science* 308, no. 5728 (2005): 1635–38.

2: Infections

18 Uninfected fruit flies live shorter lives: T. Brummel et al., "*Drosophila* Lifespan Enhancement by Exogenous Bacteria," *Proceedings of the National Academy of Sciences* 101, no. 35 (2004): 12974–79.

19 Germ-free rats need more water and food: B. S. Wostmann et al., "Dietary Intake, Energy Metabolism, and Excretory Losses of Adult Germ-Free Wistar Rats," *Laboratory Animals Science* 33 (1983): 46–50.

Reconstitution of germ-free mice makes them fat and insulin-resistant: F. Backhed et al., "The Gut Microbiota as an Environmental Factor That Regulates Fat Storage," *Proceedings of the National Academy of Sciences* 101, no. 44 (2004): 15718–23.

Composition of gut bacteria may predispose some people to obesity: ibid; C. U. Nwokolo et al., "Plasma Ghrelin Following Cure of *Helicobacter pylori*," *Gut* 52, no. 5 (2003): 637–40.

20 Mice without germs don't develop functional immune systems: L. V. Hooper and J. I. Gordon, "Commensal Host-Bacterial Relationships in the Gut," *Science* 292, no. 5519 (2001): 1115–18.

Alterations in the immune system of germ-free rabbits: K. J. Rhee et al., "Role of Commensal Bacteria in Development of Gut-Associated Lymphoid Tissues and Preimmune Antibody Repertoire," *Journal of Immunology* 172, no. 2 (2004): 1118–24; D. Lanning et al., "Intestinal Microflora and Diversification of the Rabbit Antibody Repertoire," *Journal of Immunology* 165, no. 4 (2000): 2012–19; F. Melchers and A. G. Rolink, "B-Lymphocyte Development and Biology," in *Fundamental Immunology*, edited by W. E. Paul (Philadelphia: Lippincott-Raven, 1999); J. J. Cebra, "Influences of Microbiota on Intestinal Immune System Development," *American Journal of Clinical Nutrition* 69, no. 5 (1999): 1046S–51S.

Germ-free mice get inflammatory bowel diseases: D. Kelly et al., "Commensal Anaerobic Gut Bacteria Attenuate Inflammation by Regulating Nuclear-Cytoplasmic Shuttling of PPAR-gamma and RelA," *Nature Immunology* 5, no. 1 (2004): 104–12.

Bacteria suppress allergies in mice: Zuany-Amorim, C. et al., "Suppression of Airway Eosinophilia by Killed Mycobacterium Vaccae-Induced Allergen-Specific Regulatory T-Cells," *Nature Medicine* 8, no. 6 (2002): 625–9.

21 Bacteria control human genes: L. V. Hooper et al., "Molecular Analysis of Commensal Host-Microbial Relationships in the Intestine," *Science* 291, no. 5505 (2001): 881–84; D. M. Mutch et al., "Impact of Commensal Microbiota on Murine Gastrointestinal Tract Gene Ontologies," *Physiological Genomics* 19, no. 1 (2004): 22–31.

22 The nature of bacterial biofilms: P. Stoodley et al., "Biofilms as Complex Differentiated Communities," *Annual Review of Microbiology* 56 (2002): 187–209.

Bacterial biofilms exhibit quorum sensing: M. B. Miller and B. L. Bassler, "Quorum Sensing in Bacteria," *Annual Review of Microbiology* 55 (2001): 165–99.

23 Bacterial biofilms behave like multicellular organisms: G. O'Toole, H. B. Kaplan, and R. Kolter, "Biofilm Formation as Microbial Development," *Annual Review of Microbiology* 54 (2000): 49–79; J. A. Shapiro, "Thinking About Bacterial Populations as Multicellular Organisms," *Annual Review of Microbiology* 52 (1998): 81–104.

Childhood infections suppress development of allergies and asthmas: "Hygiene Hypothesis," WGBH Educational Foundation and Clear Blue Sky Productions, www.pbs.org/wgbh/evolution/library/10/4/1_104_07.html (2001).

24   Childhood exposure to stables and farm milk limits development of allergies ands asthmas: J. Riedler et al., "Exposure to Farming in Early Life and Development of Asthma and Allergy: A Cross-Sectional Survey," *Lancet* 358, no. 9288 (2001): 1129–33.

Chinese children exhibit similar correlations between childhood exposure to infections and incidence of allergies and asthmas: G. W. Wong et al., "Factors Associated with Difference in Prevalence of Asthma in Children from Three Cities in China: Multicentre Epidemiological Survey," *BMJ* 329, no. 7464 (2004): 486.

Childhood exposure to gram-negative bacteria reduces incidence of allergies: C. Braun-Fahrlander et al., "Environmental Exposure to Endotoxin and Its Relation to Asthma in School-Age Children," *New England Journal of Medicine* 347, no. 12 (2002): 869–77.

Childhood exposure of children to gram-negative bacteria limits incidence of eczema: W. Phipatanakul et al., "Endotoxin Exposure and Eczema in the First Year of Life," *Pediatrics* 114, no. 1 (2004): 13–18.

Children fed fermented formula have better responses to polio vaccination: C. Mullié et al., "Increased Poliovirus-Specific Intestinal Antibody Response Coincides with Promotion of *Bifidobacterium longum-infantis* and *Bifidobacterium breve* in Infants: A Randomized, Double-Blind, Placebo-Controlled Trial," *Pediatric Research* 56, no. 5 (2004): 791–95.

25   Elimination of *H. pylori* from gastrointestinal tract leads to increased appetite, weight gain, and esophageal reflux: C. U. Nwokolo et al., "Plasma Ghrelin Following Cure of *Helicobacter pylori*," *Gut* 52, no. 5 (2003): 637–40.

Acute lymphoblastic leukemia accounts for 25 percent of all childhood cancers: "Childhood Acute Lymphoblastic Leukemia," www.meb.uni-bonn.de/cancer.gov/CDR0000062923.htm (2005).

26   Lack of childhood infection predisposes children to acute lymphoblastic leukemia: C. Gilham et al., "Day Care in Infancy and Risk of Childhood Acute Lymphoblastic Leukaemia: Findings from UK Case-Control Study," *BMJ* 330 (2005): 1294.

28   In the United States, the number of people with asthma will nearly double by 2020: G. Pappas et al., "Potentially Avoidable Hospitalizations: Inequalities in Rates Between US Socioeconomic Groups," *American Journal of Public Health* 87, no. 5 (1997): 811–16.

Non-Hispanic white children more likely than Hispanic children to have had allergies: "Summary Health Statistics for U.S. Children: National Health Interview Survey, 2002," in *Vital and Health Statistics*, U.S. Department of Health and Human Services, Centers for Disease Control and Prevention, National Center for Health Statistics, 2004.

29 Mitochondria are bacteria: L. Margulis and D. Sagan, *Acquiring Genomes: A Theory of the Origin of Species* (New York: Basic Books, 2002).

30 The role of mitochondria in cell life and death: X. Wang, "The Expanding Role of Mitochondria in Apoptosis," *Genes and Development* 15, no. 22 (2001): 2922–33.

3: Bugs in Our Genes

39 Human DNA contains large amounts of viral DNA, some of which represents intact active viral genomes: R. Belshaw et al., "Long-Term Reinfection of the Human Genome by Endogenous Retroviruses," *Proceedings of the National Academy of Sciences* 101, no. 14 (2004): 4894–99; F. P. Ryan, "Human Endogenous Retroviruses in Health and Disease: A Symbiotic Perspective," *Journal of the Royal Society of Medicine* 97, no. 12 (2004): 560–65.

41 Viral infection changed course of human evolution: J. F. Hughes and J. M. Coffin, "Evidence for Genomic Rearrangements Mediated by Human Endogenous Retroviruses During Primate Evolution," *Nature Genetics* 29, no. 4 (2001): 487–89; H. H. Kazazian Jr., "Mobile Elements: Drivers of Genome Evolution," *Science* 303, no. 5664 (2004): 1626–32.

Chromosomal alterations induced by viral infections changed course of human evolution: Hughes and Coffin, "Evidence for Genomic Rearrangements."

42 Brightest marsupials cannot compete with some of the dumbest placental mammals: S. J. Gould, ed., *The Book of Life* (New York: W. W. Norton, 2001).

Peg10 is derived from a viral infection and contributes to development of the placenta in mammals: R. Ono et al., "A Retrotransposon-Derived Gene, PEG10, Is a Novel Imprinted Gene Located on Human Chromosome 7q21," *Genomics* 73, no. 2 (2001): 232–37; E. M. Ostertag and H. H. Kazazian Jr., "Biology of Mammalian L1 Retrotransposons," *Annual Review of Genetics* 35 (2001): 501–38.

43 Viral infection contributed to origin of modern human Y chromosome: A. Schwartz et al., "Reconstructing Hominid Y Evolution: X-Homologous

Block, Created by X-Y Transposition, Was Disrupted by Yp Inversion Through LINE-LINE Recombination," *Human Molecular Genetics* 7, no. 1 (1998): 1–11.

Human chromosome still contains infectious viral genes: R. Belshaw et al., "Long-Term Reinfection of the Human Genome by Endogenous Retroviruses," *Proceedings of the National Academy of Sciences* 101, no. 14 (2004): 4894–99.

44 Bacterial genes in human chromosomes: Consortium, I.H.G.S., "Initial Sequencing and Analysis of the Human Genome," *Nature* 409 (2001): 860.

45 Bacterial genes in human genome apparently not ancestral genes: S. L. Salzberg et al., "Microbial Genes in the Human Genome: Lateral Transfer or Gene Loss?" *Science* 292, no. 5523 (2001): 1903–06.

46 Mitochondrial genes in human chromosomes: M. Woischnik and C. T. Moraes, "Pattern of Organization of Human Mitochondrial Pseudogenes in the Nuclear Genome," *Genome Research* 12, no. 6 (2002): 885–93.

Functional mitochondrial genes inside human chromosomes: Q. Wang et al., "Mitochondrial DNA Control Region Sequence Variation in Migraine Headache and Cyclic Vomiting Syndrome," *American Journal of Medical Genetics* 131A, no. 1 (2004): 50–58; K. L. Adams and J. D. Palmer, "Evolution of Mitochondrial Gene Content: Gene Loss and Transfer to the Nucleus," *Molecular Phylogenetics and Evolution* 29, no. 3 (2003): 380–95; U. Bergthorsson et al., "Widespread Horizontal Transfer of Mitochondrial Genes in Flowering Plants," *Nature* 424, no. 6945 (2003): 197–201; K. L. Adams et al., "Repeated, Recent and Diverse Transfers of a Mitochondrial Gene to the Nucleus in Flowering Plants," *Nature* 408, no. 6810 (2000): 354–57.

Role of mitochondrial genes in human illness: Wang, "Mitochondrial DNA Control Region Sequence Variation"; V. A. McKusick, "Kearns-Sayre Syndrome," www3.ncbi.nlm.nih.gov/entrez/dis;pomim.cgi?id=530000 (1992); P. Riordan-Eva, "Neuro-ophthalmology of Mitochondrial Diseases," *Current Opinions in Ophthalmology* 11, no. 6 (2000): 408–12; P. Riordan-Eva et al., "The Clinical Features of Leber's Hereditary Optic Neuropathy Defined by the Presence of a Pathogenic Mitochondrial DNA Mutation," *Brain* 118, pt. 2 (1995): 319–37; V. A. McKusick, "Myoclonic Epilepsy Associated with Ragged-Red Fibers," www3.ncbi.nlm.nih.gov/entrez/dispomim.cgi?id=545000 (1992); J. Lauber et al., "Mutations in Mitochondrial tRNA Genes: A Frequent Cause of Neuromuscular Diseases," *Nucleic Acids Research* 19, no. 7 (1991): 1393–37; S. Sundaram, "Cyclic Vomiting Syndrome," www.emedicine.com/ped/topic2910.htm (2002).

4: Sepsis and Self-Realization

49 David Vetter, the "Boy in the Bubble": S. McVicker, "Bursting the Bubble," *Houston Press*, April 10, 1997.

54 Pathogenesis of Crohn's disease: A. L. Hart et al., "The Role of the Gut Flora in Health and Disease, and Its Modification as Therapy," *Alimentary Pharmacology & Therapeutics* 16, no. 8 (2002): 1383–93; F. Shanahan, "Crohn's Disease," *Lancet* 359, no. 9300 (2002): 62–69.

55 Rheumatoid arthritis and gut flora: M. Malin et al., "Increased Bacterial Urease Activity in Faeces in Juvenile Chronic Arthritis: Evidence of Altered Intestinal Microflora?" *British Journal of Rheumatology* 35, no. 7 (1996): 689–94.

58 Administration of *Lactobacilli* helps treat rotaviral infections: E. Isolauri, P. V. Kirjavainen, and S. Salminen, "Probiotics: A Role in the Treatment of Intestinal Infection and Inflammation?" *Gut* 50, supp. 3 (2002): iii54–iii59.

Eating *Lactobacilli* appears to improve prognosis of patients with Crohn's disease: M. Malin et al., "Promotion of IgA Immune Response in Patients with Crohn's Disease by Oral Bacteriotherapy with *Lactobacillus GG*," *Annals of Nutrition and Metabolism* 40, no. 3 (1996): 137–45.

59 Pig whipworms help treat patients with Crohn's disease: I. Wickelgren, "Immunotherapy: Wielding Worms at Asthma and Autoimmunity," *Science* 305, no. 5681 (2004): 171; I. Wickelgren, "Immunotherapy: Can Worms Tame the Immune System?" *Science* 305, no. 5681 (2004): 170–71.

Infection of mice with parasitic worms renders mice allergy-free: Wickelgren, "Wielding Worms at Asthma and Autoimmunity" and "Can Worms Tame the Immune System?"

60 Eating worm eggs helps treat mice with MS-like disease: Wickelgren, "Wielding Worms at Asthma and Autoimmunity" and "Can Worms Tame the Immune System?"

61 David Vetter: McVicker, "Bursting the Bubble."

5: The Dark Side

70 World Health Organization comments on the future of infectious disease: *Removing Obstacles to Healthy Development*, World Health Organization

Report on Infectious Diseases, www.who.int/infectious-disease-report/pages/textonly.html (1999).

76 John Keats's tuberculosis: H. F. Dowling, *Fighting Infection: Conquests of the Twentieth Century* (Cambridge, MA: Harvard University Press, 1977).

Tuberculosis "a disease of the young, pure, and passionate": "Tuberculosis: Ancient Enemy, Present Threat," www2.niaid.nih.gov/Newsroom/Focus On/tb02/story.htm (2004).

Koch's discovery of mycobacteria as causative agent of tuberculosis: R. Guyer, "Tuberculosis: Out of Control Again," http://science-education.nih.gov/nihHTML/ose/snapshots/multimedia/ritn/Tuberculosis/tb1.html (1994).

78 Tuberculosis a thing of the past: S. Sontag, *Illness as Metaphor* (New York: Farrar, Straus, and Giroux, 1978).

History of tuberculosis: Guyer, "Tuberculosis: Out of Control Again."

Deaths from tuberculosis: V. Batra and J. Y. Ang, "Tuberculosis," www.emedicine.com/ped/topic2321.htm (2004).

79 Malaria and the fall of Rome: A. Thompson, "Malaria and the Fall of Rome," www.bbc.co.uk/history/ancient/romans/malaria_01.shtml (2005).

Life cycle of malaria: "Frequently Asked Questions About Malaria," www.cdc.gov/malaria/faq.htm (2004).

81 Epidemiology of measles: "Epidemiology of Measles," *Morbidity and Mortality Weekly* 53, no. 31 (2004): 713–16.

82 Life of Pasteur: R. Vallery-Radot, *Life of Pasteur* (New York: Doubleday, Page, 1923).

6: Taking a Turn for the Worse

85 Shift in funding from infectious disease to chronic disease research: A. S. Fauci, "Infectious Diseases: Considerations for the 21st Century," *Clinical Infectious Diseases* 32, no. 5 (2001): 675–85; L. Garrett, *The Coming Plague* (New York: Farrar, Straus, and Giroux, 1994).

"Health for All 2000" accord: D. Noah and G. Fidas, "The Global Infectious Disease Threat and Its Implications for the United States," www.cia.gov/cia/reports/nie/report/nie99-17d.html (2000).

86 Epidemiology of polio: M. S. Smolinski, M. A. Hamburg, and J. Lederberg, eds., *Microbial Threats to Health: Emergence, Detection, and Response* (Washington, DC: National Academies Press, 2003).

Pathogenic microbes identified since 1973: Noah and Fidas, "Global Infectious Disease Threat."

88 Long-lived aspen trees: G. N. Callahan, *Faith, Madness, and Spontaneous Human Combustion* (New York: St. Martin's Press, 2002).

90 Global infectious disease threat: Noah and Fidas, "Global Infectious Disease Threat."

91 Persistence of infection depends on population density: A. Tran et al., "Mapping Disease Incidence in Suburban Areas Using Remotely Sensed Data," *American Journal of Epidemiology* 156, no. 7 (2002): 662–68.

Displaced people in the world: Noah and Fidas, "Global Infectious Disease Threat."

92 Multidrug-resistant *Streptococci* and their spread from Spain to rest of the world: ibid.

98 How viruses avoid immune responses: D. Tortorella et al., "Viral Subversion of the Immune System," *Annual Review of Immunology* 18 (2000): 861–926.

Thirty percent of people continuously infected with *S. aureus*: T. Herchline, "Staphylococcal infections," www.emedicine.com/med/topic2166.htm (2004).

99 Antibiotic-resistant strains of staph: E. Askari, "Germs Develop a Deadly Defense," www.freep.com/news/health/nstaph12_20021112.htm (2002).

100 By 2000, more than 55 percent of staph isolated from critical-care patients resistant to methicillin: Smolinski et al., *Microbial Threats to Health*.

Vancomycin-resistant staph in Michigan: Askari, "Germs Develop a Deadly Defense."

102 Factors contributing to infectious disease: Noah and Fidas, "Global Infectious Disease Threat."

7: The Occult

107 MS cluster in Galion, Ohio: National MS Society, *Information Sourcebook: Clusters*, www.nationalmssociety.org/pdf/sourcebook/clusters.pdf (2004).

108 Demyelinating diseases that follow vaccination and flare-ups of MS that follow viral infections: R. T. Johnson, "Demyelinating Diseases," in *The Infectious Etiology of Chronic Diseases*, edited by S. L. Knobler, S. O'Con-

nor, S. M. Lemon, and M. Najafi (Washington, DC: National Academies Press, 2005).

MS patients have higher reactivity to several viruses: ibid.

109 Characteristics of varicella virus and Epstein-Barr virus: ibid.

110 Epidemiology of cervical cancer: E. L. Franco, "The Role of Viruses in Oncogenesis: Human Papilloma Viruses and Cervical Cancer as a Paradigm," in Knobler et al., *Infectious Etiology of Chronic Diseases*.

111 Epidemiology of hepatitis B virus: W. Mason, "Chronic Hepatitis B Infections," in Knobler et al., *Infectious Etiology of Chronic Diseases*.

112 Epidemiology of hepatitis C infection: A. Kamal, "Progression of Hepatitis C Virus Infection with and Without Schistosomiasis," in Knobler et al., *Infectious Etiology of Chronic Diseases*; R. B. Belshe, "The Origins of Pandemic Influenza—Lessons from the 1918 Virus," *New England Journal of Medicine* 353, no. 21 (2005): 2209–11; WHO, *The World Health Report 1996: Fighting Disease, Fostering Development* (Geneva: World Health Organization, 1996); D. Wolf, "Hepatitis, Viral," www.emedicine.com/med/topic3180.htm (2005).

Epidemiology of schistosomiasis: D. Engels et al., "The Global Epidemiological Situation of Schistosomiasis and New Approaches to Control and Research," *Acta Tropica* 82, no. 2 (2002): 139–46; L. Chitsulo et al., "The Global Status of Schistosomiasis and Its Control," *Acta Tropica* 77, no. 1 (2000): 41–51.

113 Combined effects of *Schistosoma mansoni* and hepatitis C infections: A. Kamal, "Progression of Hepatitis C Virus Infection"; S. Kamal et al., "Clinical, Virological and Histopathological Features: Long-Term Follow-up in Patients with Chronic Hepatitis C Co-infected with *S. mansoni*," *Liver* 20, no. 4 (2000): 281–89.

World Health Organization estimates of cancers that could be prevented if relevant infectious diseases were controlled: WHO, *Fighting Disease, Fostering Development*.

114 *H. Pylori* and peptic ulcers: "*H. Pylori* and Peptic Ulcer," http://digestive.niddk.nih.gov/ddiseases/pubs/hpylori/ (2004).

115 Chlamydia and atherosclerosis: M. Dunne, "Infectious Agents and Cardiovascular Disease," in Knobler et al., *Infectious Etiology of Chronic Dis-*

NOTES • 265

eases; E. F. Torrey and R. H. Yolken, "*Toxoplasma gondii* and Schizophrenia," *Emerging Infectious Diseases* 9, no. 11 (2003): 1375–80; Smolinski et al., *Microbial Threats to Health*.

116   Infection and epilepsy: J. W. Sander, "Infectious Agents and Epilepsy," in Knobler et al., *Infectious Etiology of Chronic Diseases*.

117   Bacteria, viruses, and worms associated with chronic diseases: Knobler et al., *Infectious Etiology of Chronic Diseases*.

8: The Truth About Insanity

120   Deinstitutionalization of the mentally ill in the United States: "Many Americans with Untreated Illnesses Have Nowhere to Go," www.psychlaws.org/GeneralResources/fact11.htm (2003).

121   Epidemiology of schizophrenia: "A Short Introduction to Schizophrenia," www.schizophrenia.com/family/schizintro.html (2005); "Heredity and Genetics of Schizophrenia," www.schizophrenia.com/research/hereditygen.htm (2005).

Genes active in schizophrenic brains: Yolken, *Infectious Agents and Schizophrenia*; F. Yee and R. H. Yolken, "Identification of Differentially Expressed RNA Transcripts in Neuropsychiatric Disorders," *Biological Psychiatry* 41, no. 7 (1997): 759–61.

Effects of clozapine on retroviruses and endogenous retroviruses active during fetal brain development: R. H. Yolken et al., "Endogenous Retroviruses and Schizophrenia," *Brain Research Review* 31, no. 2–3 (2000): 193–99.

123   Epidemiology of *T. gondii*: J. L. Jones, D. Kruszon-Moran, and M. Wilson, "*Toxoplasma gondii* Infection in the United States, 1999–2000," *Emerging Infectious Diseases* 9, no. 11 (2003): 1371–74.

Breast-fed children more likely to develop schizophrenia: E. F. Torrey, R. Rawlings, and R. H. Yolken, "The Antecedents of Psychoses: A Case-Control Study of Selected Risk Factors," *Schizophrenia Research* 46, no. 1 (2000): 17–23.

Schizophrenic patients infected with *T. gondii* show greater cognitive impairment than people with schizophrenia alone: E. F. Torrey and R. H. Yolken, "*Toxoplasma gondii* and Schizophrenia," *Emerging Infectious Diseases* 9, no. 11 (November 2003), www.cdc.gov/ncidod/EID/vol9no11/03-0143.htm.

Effects of *T. gondii* infection on humans: C. Zimmer, "Parasites Make Scaredy-Rats Foolhardy," *Science* 289 (2000): 525.

124 *Journal of the American Medical Association* quote: L. A. Snider and S. E. Swedo, "Pediatric Obsessive-Compulsive Disorder," *JAMA* 284, no. 24 (2000): 3104–06.

125 Obsessive-compulsive disorder and streptococcal infections: S. E. Swedo, H. L. Leonard, and J. L. Rapoport, "The Pediatric Autoimmune Neuropsychiatric Disorders Associated with Streptococcal Infection (PANDAS) Subgroup: Separating Fact from Fiction," *Pediatrics* 113, no. 4 (2004): 907–11.

126 Obsessive-compulsive and tic disorders associated with streptococcal infections: S. E. Swedo et al., "Pediatric Autoimmune Neuropsychiatric Disorders Associated with Streptococcal Infections: Clinical Description of the First 50 Cases," *American Journal of Psychiatry* 155, no. 2 (1998): 264–71.

Complete plasma exchange and intravenous administration of mixed human immunoglobulins for treatment of PANDAS: S. J. Perlmutter et al., Therapeutic Plasma Exchange and Intravenous Immunoglobulin for Obsessive-Compulsive Disorder and Tic Disorders in Childhood," *Lancet* 354, no. 9185 (1999): 1153–58.

127 Involvement of autoantibodies in pathogenesis of PANDAS: Swedo et al., "Pediatric Autoimmune Neuropsychiatric Disorders Associated with Streptococcal Infection (PANDAS) Subgroup"; A. J. Church, R. C. Dale, and G. Giovannoni, "Anti-Basal Ganglia Antibodies: A Possible Diagnostic Utility in Idiopathic Movement Disorders?" *Archives of Disease in Childhood* 89, no. 7 (2004): 611–14; R. C. Dale and I. Heyman, "Post-Streptococcal Autoimmune Psychiatric and Movement Disorders in Children," *British Journal of Psychiatry* 181 (2002): 188–90; S. E. Swedo and P. J. Grant, "Annotation: PANDAS: A Model for Human Autoimmune Disease," *Journal of Child Psychology and Psychiatry* 46, no. 3 (2005): 227–34; L. A. Snider and S. E. Swedo, "Post-Streptococcal Autoimmune Disorders of the Central Nervous System," *Current Opinions in Neurology* 16, no. 3 (2003): 359–65.

128 Effects of parasitic infections on behavior of ants: reviewed in G. N. Callahan, "Madness," *Emerging Infectious Diseases* 8, no. 9 (2002): 998–1002.

129 Effects of parasitic infections on fish behavior: R.P.E. Yanong, "Nematode (Roundworm) Infection in Fish," http://edis.ifas.ufl.edu/FA091 (2002).

Effects of viral infections on human behavior: R. Hunt, "Microbiology and Immunology on Line," http://pathmicro.med.sc.edu/virol/herpes.htm (2004).

9: Red Dawn

134 2000 Central Intelligence Agency report: D. Gordon and D. Noah, "The Global Infectious Disease Threat and Its Implications for the United States," www.cia.gov/cia/reports/nie/report/nie99-17d.html (2000).

135 U.S. National Academies: *Microbial Threats to Health. Emergence, Detection, and Response* (Washington, DC: National Academies Press, 2003).

Epidemiology of hepatitis C: ibid.

136 Likely resurgence of tuberculosis: ibid.

Epidemiology of tuberculosis: National Center for HIV, STD, and TB Prevention, Division of Tuberculosis Elimination, www.cdc.gov/nchstp/tb/surv/surv2004/default.htm (2005).

Deaths from hospital-acquired infections: Gordon and Noah, "The Global Infectious Disease Threat."

136–
37 Infectious diseases in the United States and the world: ibid.

139 Enterovirus 71 epidemiology: B. A. Brown et al., "Molecular Epidemiology and Evolution of Enterovirus 71 Strains Isolated from 1970 to 1998," *Journal of Virology* 73, no. 12 (1999): 9969–75; J. P. Alexander Jr. et al., "Enterovirus 71 Infections and Neurologic Disease—United States, 1977–1991," *Journal of Infectious Diseases* 169, no. 4 (1994): 905–08; T. N. Wu et al., "Sentinel Surveillance for Enterovirus 71, Taiwan, 1998," *Emerging Infectious Diseases* 5, no. 3 (1999): 458–60.

Marburg hemorrhagic fever outbreak in Angola: "Marburg Virus Death Toll Hits 180," www.cnn.com/2005/HEALTH/conditions/04/08/angola.marbug (2005); E. Marris, "Marburg Workers Battle to Win Trust of Locals," *Nature* 434 (2005): 946.

Antibiotic-resistant strains of *Staphylococcus aureus*, malaria, and tuberculosis: Centers for Disease Control and Prevention, Information by Emerging or Reemerging Infectious Disease Topic, www.cdc.gov/ncidod/diseases/eid/disease_sites.ht (2005).

140 United Nations report: *Environmental Changes Are Spreading Infectious Diseases*, www.un.org/apps/news/story.aspNewsID=13407&Cr=infectious&Cr1=disease# (2005).

141 Effects of environmental changes on disease spread and human susceptibility: "Scientists Link Environment with Infectious Disease Spread,"
www.scienceinafrica.co.za/2005/march/envirodisease.htm (2005).

Annual deaths from malaria: "Malaria Facts," www.cdc.gov/malaria/
facts.htm (2004).

142 National Institute of Allergy and Infectious Diseases warning about impact
of environmental changes on infectious diseases: "Research on Ecologic
and Environmental Factors Influencing Emergence," www.niaid.nih.
gov/publications/execsum/1a.htm (2001).

Northridge earthquake and valley fever: "*Coccidioidomycosis* Following the
Northridge Earthquake—California," *Morbidity and Mortality Weekly* 43,
no. 10 (1994): 194B195, 421B423.

Dam construction and schistosomiasis outbreaks: J. Lederberg, R. E. Shope,
and S. C. Oaks, eds., *Emerging Infections: Microbial Threats to Health in the
United States* (Washington, DC: National Academies Press, 1992).

Deforestation and malaria: Pan American Health Organization, "Malaria
in the Americas," *Epidemiology Bulletin* 13 (1992): 1B6.

143 Lyme disease in America: Lederberg et al., *Emerging Infections.*

144 Northern Prairie Wildlife Center and reports of malformed frogs: I. K. Lo-
effler et al., "Leaping Lopsided: A Review of the Current Hypotheses Regarding Etiologies of Limb Malformations in Frogs," *Anatomical Record*
265, no. 5 (2001): 228–45.

Trematode infections and frog malformations: J. Kaiser, "A Trematode Parasite Causes Some Frog Deformities," *Science* 284, no. 5415 (1999): 731,
733; P. T. Johnson et al., "Review of the Trematode Genus *Ribeiroia
(Psilostomidae)*: Ecology, Life History and Pathogenesis with Special Emphasis on the Amphibian Malformation Problem," *Advances in Parasitol-
ogy* 57 (2004): 191–253; J. Kaiser, "Ecology: Fifty Years of Deformed
Frogs," *Science* 301, no. 5635 (2003): 904.

145 Effects of PCBs on marine mammals' immune systems: Garrett, *Coming
Plague* (New York: Farrar, Strauss & Giroux, 1994).

10: The Spider in Room 911
154 Cluster of cases of SARS in Hong Kong: K. W. Tsang et al., "A Cluster of
Cases of Severe Acute Respiratory Syndrome in Hong Kong," *New England Journal of Medicine* 348, no. 20 (2003): 1977–85.

WHO meeting about SARS in Vietnam: B. Reilley et al., "SARS and Carlo Urbani," *New England Journal of Medicine* 348, no. 20 (2003): 1951–52.

155 WHO and CDC issue travel advisories about SARS: "SARS: Timeline of an Outbreak," http://my.webmd.com/content/article/63/72068.htm (2005).

Singapore and Thailand report cases of SARS: "Update: Outbreak of Severe Acute Respiratory Syndrome — Worldwide, 2003," *Morbidity and Mortality Weekly Report* 52 (2003): 269–72.

156 Public Health Agency of Canada reports eleven cases of SARS: "Summary of Severe Acute Respiratory Syndrome (SARS) Cases: Canada and International," www.phac-aspc.gc.ca/sars-sras/eu-ae/sars20030323_e.html#tab1 (2003).

Vietnamese Ministry of Health, Hong Kong Department of Health, and Taiwan Department of Public Health report probable cases of SARS: "Outbreak of Severe Acute Respiratory Syndrome — Worldwide, 2003."

157 Centers for Disease Control announces fifty-one reports of suspected cases of SARS in the United States: ibid.

159 Centers for Disease Control announces completion of full-length genetic sequence of the SARS-CoV RNA: "SARS-Associated Coronavirus (SARS-CoV) Sequencing," www.cdc.gov/ncidod/sars/sequence.htm (2003).

World Health Organization announces coronavirus as definitive cause of SARS: D. M. Skowronski et al., "Severe Acute Respiratory Syndrome (SARS): A Year in Review," *Annual Review of Medicine* 56 (2005): 357–81.

161 Epidemiology of SARS: ibid.

162 *New York Times* reports SARS is over: J. Yardley, "After Its Epidemic Arrival, SARS Vanishes," *New York Times*, May 15, 2005.

11: Diseases on the Fly

163 Wayne and Sandra Trowbridge: A. P. Hundley, "Six Months Later, Milliken Man Loses West Nile Fight," www.greeleytrib.com/apps/pbcs.dll/article?AID=/20040218/NEWS/102180013&rs=1&template=printart (2004).

164 West Nile Virus infections in New York City: D. S. Asni et al., "The West Nile Virus Outbreak of 1999 in New York: The Flushing Hospital Experience," *Clinical Infectious Diseases* 30, no. 3 (2000): 413–18.

165 West Nile virus and corvids: "West Nile Virus Activity—United States 2001," *Morbidity and Mortality Weekly* 51 (2002): 497–501; G. L. Campbell et al., "West Nile Virus," *Lancet Infectious Diseases* 2, no. 9 (2002): 519–29.

167 Patricia Heller and West Nile virus infection: L. Neergaard, "W. Nile Thwarts Young, Healthy," *Fort Collins Coloradoan*, May 31, 2005.

168 West Nile virus statistics 2002: E. B. Hayes, "Epidemiology and Transmission Dynamics of West Nile Virus Disease," *Emerging Infectious Diseases* 11, no. 8 (2005): 1167–73.

169 West Nile virus statistics 2003–2005: "Update West Nile Virus Activity—United States 2005," *Morbidity and Mortality Weekly* 54 (2005): 877–78.

170 Cases of paralysis from West Nile virus: Neergaard, "W. Nile Thwarts Young, Healthy."

171 Malaria in the United States. R. Desowitz, *The Malaria Capers* (New York: W. W. Norton, 1993).

172 Modern outbreaks of malaria in United States: J. R. Zucker, "Changing Patterns of Autochthonous Malaria Transmission in the United States: A Review of Recent Outbreaks," *Emerging Infectious Diseases* 2, no. 1 (1996): 37–43.

173 Dracunculiasis: "Fact Sheet: Dracunculiasis," www.cdc.gov/ncidod/dpd/parasites/dracunculiasis/factsht_dracunculiasis.htm (2004).

174 Pathogenesis of malaria: L. H. Miller, M. F. Good, and G. Milon, "Malaria Pathogenesis," *Science* 264, no. 5167 (1994): 1878–83.

Malaria kills more people than HIV, measles, tuberculosis, and leprosy combined: "Malaria in Southern Africa," www.malaria.org.za/Malaria_Risk/General_Information/general_information.html (2003).

175 Malaria in Ecuador and Bolivia: P. Driessen, "Double Standards on Disease Control: Anti-Pesticide Policies Violate Human Rights and Condemn Millions to Needless Death," *Canada Free Press*, May 20, 2005, Toronto.

Focused spraying of DDT has no effect on people or environment: *Microbial Threats to Health. Emergence, Detection, and Response* (Washington, DC: National Academies Press, 2003).

176 Cyril Boynes on effects of preventing use of DDT in Africa: C. Boynes, "EU Position on DDT Violates Human Rights," *Canada Free Press*, February 16, 2005.

Cryptic malaria in United States: Zucker, "Changing Patterns of Autochthonous Malaria Transmission in the United States."

178 1981 outbreak of dengue fever: M. G. Guzman, "Global Voices of Science. Deciphering Dengue: The Cuban Experience," *Science* 309, no. 5740 (2005): 1495–97; G. Kouri et al., "Reemergence of Dengue in Cuba: A 1997 Epidemic in Santiago de Cuba," *Emerging Infectious Diseases* 4, no. 1 (1998): 89–92.

Dengue fever in Cuba: M. Rodriguez-Roche et al., "Dengue Virus Type 3: Cuba 2000–2002," *Emerging Infectious Diseases* 11, no. 5 (2005).

179 Dengue-3 mutant in East Africa, Sri Lanka, and Central America: "Study Traces Global Spread of Virulent Dengue Virus to U.S. Doorstep," www.sciencedaily.com/releases/2003/06/030626235349.htm (2003).

Genetics and the outcome of dengue virus infection: H. A. Stephens et al., "HLA-A and -B Allele Associations with Secondary Dengue Virus Infections Correlate with Disease Severity and the Infecting Viral Serotype in Ethnic Thais," *Tissue Antigens* 60, no. 4 (2002): 309–18.

180 Centers for Disease Control: *Microbial Threats to Health.*

181 Recent spread of dengue virus: ibid.

Dengue now endemic in more than 100 countries: "Dengue Fever," *Medical Encyclopedia*, www.nlm.nih.gov/medlineplus/ency/article/001373.htm/ Causes,%20incidenice,%20and%20risk%20factors (2004).

Epidemiology of dengue: "Dengue and Dengue Hemorrhagic Fever," www.who.int/mediacentre/factsheets/fs117/en/print.html (2002).

Dengue fever in Hawaii: P. V. Effler et al., "Dengue Fever, Hawaii, 2001–2002," *Emerging Infectious Diseases* 11, no. 5 (2005): 742–49.

182 Centers for Disease Control: *Microbial Threats to Health.*

12: Agents of Change
184 Maureen and Robert Stevens and anthrax: "Anthrax Mystery Still Unsolved," www.cbsnews.com/stories/2003/02/25/earlyshow/health/main541988.shtml (February 26, 2003).

Johanna Huden and anthrax: E. Lipton and K. Johnson, "Tracking Bioterror's Tangled Course," *New York Times*, December 26, 2001.

185 Ottie Lundgren and anthrax: "Epidemiologic Information on Bioterrorism. 2004 and American Anthrax Outbreak of 2001," www.ph.ucla.edu/epi/bioter/detect/antdetect_letters.html and www.ph.ucla.edu/epi/bioter/bioterrorism.html#Anthrax (2004).

187 British experiments with anthrax on Gruinard: "Epidemiologic Information on Bioterrorism," www.ph.ucla.edu/epi/bioter/bioterrorism.html#Anthrax (2004); R. J. Manchee et al., "Decontamination of *Bacillus anthracis* on Gruinard Island?" *Nature* 303, no. 5914 (1983): 239–40; "Britain's Anthrax Island," http://news.bbc.co.uk/1/hi/scotland/1457035.stm (2001).

188 UCLA School of Epidemiology: www.ph.ucla.edu/epi/bioter/detect/antdetect_letters.html.

190 Siege at Caffa: M. Wheelis, "Biological Warfare at the 1346 Siege of Caffa," *Emerging Infectious Diseases* 8, no. 9 (2002).

193 Plague pathogenesis: G. R. Cornelis, "*Yersinia* Type III Secretion: Send in the Effectors," *Journal of Cell Biology* 158, no. 3 (2002): 401–08.

194 Select List of biologic agents: "Select Agents Program," www.cdc.gov/od/sap.

195 Terrorist uses of ricin: "Ricin," www.emedicinehealth.com/articles/42387-1.asp (2006).

13: Eating Your Brains Out
199– History of kuru, scrapie, and CJD: P. Brown and R. Bradley, "1755 and All
203 That: A Historical Primer of Transmissible Spongiform Encephalopathy," *BMJ* 317, no. 7174 (1998): 1688–92.

205 Prion change similar to change from curtain to Venetian blind: ibid.

Pathogenesis of transmissible spongiform encephalopathies: "Creutzfedt-Jakob Disease," www.intelihealth.com/IH/ihtIH/WSIHW000/9339/9768.html (2003).

Epidemiology of BSE and CJD: "Bovine Spongiform Encephalopathy and Creutzfeldt-Jakob Disease," www.cdc.gov/ncidod/diseases/cjd/cjd_fact_sheet.htm (2004).

206 Rendering of animal carcasses: P. Brown et al., "Bovine Spongiform Encephalopathy and Variant Creutzfeldt-Jakob Disease: Background, Evolution, and Current Concerns," *Emerging Infectious Diseases* 7, no. 1 (2001): 6–16.

207 Origin of spongiform encephalopathies in wild cats: ibid.

208 Differences between CJD and vCJD: ibid.

Characteristics of vCJD: R. G. Will et al., "A New Variant of Creutzfeldt-Jakob Disease in the UK," *Lancet* 347, no. 9006 (1996): 921–25.

Epidemiology of BSE and vCJD: ibid.

vCJD deaths outside of Britain: "Bovine Spongiform Encephalopathy and Creutzfeldt-Jakob Disease."

209 Human origin of mad cow disease: A. C. Colchester and N. T. Colchester, "The Origin of Bovine Spongiform Encephalopathy: The Human Prion Disease Hypothesis," *Lancet* 366, no. 9488 (2005): 856–61.

211 U.S. regulations regarding slaughter of domestic livestock: "Prohibition of the Use of Certain Stunning Devices Used to Immobilize Cattle During Slaughter," www.beefusa.org/NEWSProhibitionoftheUseofCertainStunningDevicesUsedtoImmobilizeCattleDuringSlaughter14219.aspx (2005).

Continued heartbeat and blood circulation in animals after stunning: P. Comer and P. J. Huntly, "Exposure of the Human Population to BSE Infectivity over the Course of the BSE Epidemic in Great Britain and the Impact of Changes to the Over-Thirty-Month Rule," *Journal of Risk Exposure* 7 (2004): 523–43.

Prions found in animals outside of central nervous system: C. Weissmann and A. Aguzzi, "Approaches to Therapy of Prion Diseases," *Annual Review of Medicine* 56 (2005): 321–44; P. J. Bosque et al., "Prions in Skeletal Muscle," *Proceedings of the National Academy of Sciences* 99, no. 6 (2002): 3812–17; "Prions in Skeletal Muscle of Deer with Chronic Wasting Disease," *Science*, published online, January 26, 2006.

USDA study of risks to United States from BSE: J. Cohen et al., "Evaluation of the Potential for Bovine Spongiform Encephalopathy in the U.S.," www.hcra.harvard.edu/pdf/madcow.pdf (2003).

212 Advanced meat recovery process: AMI fact sheet, "Meat Derived by Advanced Meat Recovery," www.amif.org/FactSheetAdvancedMeatRecovery.pdf (2002).

Harvard-Tuskegee report about AMR and the risk of meat contamination with spinal cord: Cohen et al., "Evaluation of the Potential for Bovine Spongiform Encephalopathy in the U.S."

Most AMR meat ends in ground meat, meatballs, and taco fillings: "Meat Derived by Advanced Meat Recovery."

213   Peer review of Harvard-Tuskegee risk assessment: H. Frey et al., "Review of the Evaluation of the Potential for Bovine Spongiform Encephalopathy in the United States Conducted by the Harvard Center for Risk Analysis, Harvard School of Public Health & Center for Computational Epidemiology, College of Veterinary Medicine, Tuskegee University," www.aphis.usda. gov/lpa/issues/bse/BSE_Peer_Review.pdf (2002).

Oprah Winfrey show on mad cow disease: "Oprah Winfrey: Mad Cow Disease," www.ecomall.com/greenshopping/eioprah.htm (1996).

214   Many U.S. rendering plants not in compliance with voluntary bans in place in 1992: "Consumer Medical Groups Urge U.S. to Boost 'Mad Cow' Safety," www.mad-cow.org/~tom/consumer_health.html.

215   Inability to track individual animals from birth to slaughter: transcript of Tele-News Conference with Dr. John Clifford, chief veterinary officer, Animal and Plant Health Inspection Service; Dr. Stephen Sundlof, director, Center for Veterinary Medicine, Food and Drug Administration; and Dr. Bob Hillman, executive director, Texas Animal Health Commission, www.usda.gov/wps/portal/usdahome?contentidonly rue&contentid= 2005/06/0235.xml (Washington, DC, June 29, 2005).

Montana judge refuses to lift ban on Canadian cattle: D. Kravets, "Canadian Cattle Ban Lifted by U.S. Court," *Rocky Mountain News*, July 12, 2005.

U.S. Justice Department urges appeals court in Seattle to reopen our northern border to Canadian beef imports: ibid.

216   Japanese feel USDA not doing an adequate job: Associated Press, "Japan: U.S. Tests Would Have Missed Infected Animals," *Fort Collins Coloradoan*, July 16, 2005.

14: An Infectious Holocaust

217   Randy Shilts on mosquitoes and AIDS: R. Shilts, "Talking AIDS to Death," in *Best American Essays 1990*, edited by J. Kaplan and R. Atwan (New York: Ticknor and Fields, 1990).

218   Transmission of AIDS from chimps to humans: F. Gao et al., "Origin of HIV-1 in the Chimpanzee *Pan troglodytes troglodytes*," *Nature* 397, no. 6718 (1999): 436–41.

219 Characteristics of HIV-2: K. de Kock et al., "Epidemiology and Transmission of HIV-2: Why There Is No HIV-2 Pandemic?" *JAMA* 270 (1993): 3083–86.

220 Kaposi's sarcoma in gay men in New York City, 1981: K. B. Hymes et al., "Kaposi's Sarcoma in Homosexual Men—A Report of Eight Cases," *Lancet* 2, no. 8247 (1981): 598–600.

Centers for Disease Control and Prevention issues warning about PCP in gay men: "Kaposi's Sarcoma and Pneumocystis Pneumonia among Homosexual Men—New York City and California," *Morbidity and Mortality Weekly* 30, no. 4 (1981): 305–08.

221 No apparent danger to nonhomosexuals from Kaposi's sarcoma: L. Altman, "Rare Cancer Seen in 41 Homosexuals," *New York Times*, July 3, 1981.

222 Quote from Margaret Heckler: "HIV & AIDS," www.avert.org/historyi.htm (2004).

227 Women in underdeveloped countries know little about sexual intercourse, how to limit transmission of HIV, and the use of condoms: H. Marais, "AIDS Epidemic Update December 2004," www.unaids.org/wad2004/EPI update2004_html_en/epiO4_00_en.htm (2005).

229 U.S. policy on AIDS assistance: L. Rohter, "Prostitution Puts U.S. and Brazil at Odds on AIDS Funding Policy," *New York Times*, July 24, 2005.

231 Difficulties of AIDS vaccine: A. J. McMichael and T. Hanke, "HIV Vaccines 1983–2003," *Nature Medicine* 9, no. 7 (2003): 874–80.

234 HAART and its effect on HIV-infected men and women: N. L. Letvin and B. D. Walker, "Immunopathogenesis and Immunotherapy in AIDS Virus Infections," *Nature Medicine* 9, no. 7 (2003): 861–66; "Highly Active Antiretroviral Therapy Has Extended AIDS-free Survival Times for Some," www.ahrq.gov/research/sep01/901RA18.htm (2002).

235 Well-known people who have died from AIDS: "AIDS-Related Deaths," http://en.wikipedia.org/wiki/Category: AIDS-related_deaths (2005).

15: The Slate Wiper

238 1918 flu pandemic in Kansas: L. Goodson, "Pandemic," *Manhattan Mercury*, March 1, 1998.

242 Scientists recover 1918 flu from dead Inuit woman in Seward, Alaska: R. G. Webster, "1918 Spanish Influenza: The Secrets Remain Elusive," *Proceedings of the National Academy of Sciences* 96, no. 4 (1999): 1164–66.

1918 flu moved directly from birds to humans: T. M. Tumpey et al., "Characterization of the Reconstructed 1918 Spanish Influenza Pandemic Virus," *Science* 310, no. 5745 (2005): 77–80.

246   Influenza frequently moves directly from birds to humans: R. B. Belshe, "The Origins of Pandemic Influenza—Lessons from the 1918 Virus," *New England Journal of Medicine* 353, no. 21 (2005): 2209–11.

Resurrected 1918 flu extremely lethal in mice: V. Bubnoff, "The 1918 Flu Virus Resurrected," *Nature* 437 (2005): 794–95.

Among the vaccinated elderly, incidence of flu-related deaths seems to be increasing: J. Cohen, "Influenza Study Questions the Benefits of Vaccinating the Elderly," *Science* 307, no. 5712 (2005): 1026.

248   Effects of H5N1 flu on chickens: K. Greenfield, "On High Alert," *Time Asia*, January 26, 2004.

Chinese cases of human H5N1 influenza may have gone unreported: M. Enserink, "Talk on 'Underground' Bird Flu Deaths Rattles Experts," *Science* 310 (2005): 1409.

249   Epidemic outbreaks of H5N1 influenza: "Key Facts About Avian Influenza (Bird Virus) and Avian Influenza A (H5N1 virus)," www.cdc.gov/flu/avian/gen-info/facts.htm (2005); "Geographical Spread of H5N1 Influenza in Birds—Update 28," www.who.int/csr/don/2005_08_18/en/print.html (2005).

Deaths of bar-headed geese in north-central China: "Geographical Spread of H5N1 Influenza in Birds—Update 28"; D. Brown, "Deadly Flu Strain Shows up in Migratory Birds," *Washington Post*, July 7, 2005; D. Normile, "Avian Influenza: Potentially More Lethal Variant Hits Migratory Birds in China," *Science* 309, no. 5732 (2005): 231.

250   Effects of H5N1 avian influenza in Africa: T. Simonite, "Migrations Threaten to Send Flu South," *Nature* 437 (2005): 1212–13.

Outbreaks of H5N1 avian influenza in Turkey in 2006: M. Enserink, "At Least 14 H5N1 Infections in Turkey," http://sciencenow.sciencemag.org/cgi/content/full/2006/109/1?etoc(2006).

H5N1 avian influenza in house cats and tigers: R. Thanawongnuwech et al., "Probable Tiger-to-Tiger Transmission of Avian Influenza H5N1," *Emerging Infectious Diseases* 11, no. 5 (2005): 699–701.

H5N1 influenza in crested hawk eagles in Belgium: "Bird Flu Sparks Belgian Manhunt," http://news.bbc.co.uk/1/hi/world/europe/3948597.stm (2004).

H5N1 influenza found in pigs in China: "Avian Influenza Virus H5N1 Detected in Pigs in China," www.who.int/csr/don/2004_08_20/en/ (2005).

251 Possible human-to-human transmission of H5N1 influenza: "H5N1 Spreading Among Humans?" www.the-scientist.com/news/20050520/01 (2005); S. J. Olsen et al., "Family Clustering of Avian Influenza A (H5N1)," *Emerging Infectious Diseases* 11, no. 11 (2005).

Chinese block influenza investigations: "Chinese Rules Stymie Flu Scientist," www.the-scientist.com/news/20050706/01 (2005).

Changes in pathogenicity of H5N1 influenza: "H5N1 Spreading Among Humans?"

H5N1 avian flu resistant to Tamiflu: Q. M. Le et al., "Avian Flu: Isolation of Drug-Resistant H5N1 Virus," *Nature* 437, no. 7062 (2005): 1108.

252 Anthony Fauci announces plans to acquire flu vaccine: D. C. Page, "Avian Flu Vaccine Set for U.S. Production," *San Francisco Chronicle*, August 7, 2005.

Index •:•

asthma, 23–24, 55, 56
  increase in, 27–28
  and intestinal worm treatment, 59
atherosclerosis, 115
autoimmune diseases, 55–56, 105–6
  treatment of, 60
autoimmunity, 126–27
avian flu. *See* influenza
AZT, 234

Bacillus anthracis, 187–88, 196
bacteria
  ability to bypass immune system, 97
  abundance and importance of, 6, 14
  cells, in the human body, 8
  controlling some human genes, 21
  disease producing and non-disease
    producing, 70
  DNA of, 22
  permanent vs. transient, 12–14
  reliance of living things on, 18–19
  resistance to antibiotics, 99–101
  sites of, in human body, 10
  speed of evolution of, 101
bacterial flora, 11–12
  normal, 12–13
bacterial genes, in human genome, from
    evolutionarily recent bacterial infection,
    10–11, 44–45
Bacteroides, 11
beef. *See* cattle
Bifidobacteria, 9–10, 11, 25
bilharzia. *See* schistosomiasis
biofilms, 22–23
biosafety level (BSL), 197
bioterrorism, 185, 187–91, 194–98
  research against, 197–98
biowarfare, 190–91
bipolar disorder, 120
birds
  influenza's course of infection in, 248–49
  and West Nile virus, 164–66, 168–69
birth canal, bacteria in, 8–9
bites, diseases spread by, 67
black bane, 184
black plague. *See* bubonic plague
blood, bacterial infection of, 99
blood transfusions, and AIDS, 221, 225
Bordetella pertussis, 72

Borrelia burgdorferi, 87, 139–40
Botulinum toxin, 196
botulism, 14
bovine spongiform encephalopathies (BSE)
    (mad cow disease), 129
  link with vCJD, 208
  potential for outbreak in U.S., 211–13
  prion infection and, 210–16
  in United Kingdom, 204, 206–10
Boynes, Cyril, Jr., 176
brain, tumors in, 225
Brazil, AIDS program of, 229–30
breakbone fever. *See* dengue fever
breast-feeding
  and bacterial flora of babies, 11
  and schizophrenia, 123
bronchi, microbes in, 71
brucellosis, 27
buboes (swollen lymph nodes), 192
bubonic plague, 14, 190–92

Campylobacter jejuni, 75
Canadian beef, 214–16
  ban on importation of, to U.S., 214–15
Canadian Cattlemen's Association, 216
cancer, 109–12
  infectious diseases and, 113
Candida albicans, 9, 10, 12, 224–25
cannibalism, and kuru, 199–203
Carson, Rachel, 175
case histories
  anthrax, 183–85
  cannibalism, and kuru, 199–203
  5-ARD (girl becoming a boy), 33–37
  gas gangrene, xi–xiv
  influenza, 236–40
  mental illness, 118–19
  multiple sclerosis, 104–5
  mutant frogs, 131–33
  rabies, 65–66, 82
  rotavirus, 56–57
  SARS, 149–57
  SCID, 49–50, 61
  syphilis, 3–5
  West Nile virus, 163–64
cats, 123, 127
cattle feed, animal products in, 206–10,
    213–16
cell death, balance with cell division, 30

schizophrenia, 120–24
scrapie, 202–3, 206, 207, 209
 prion of, 210
seals, 145
"Select List" of bioterrorism agents, 194,
 196–97
semen, and AIDS, 93, 226
severe-acute respiratory syndrome (SARS), 88,
 92, 138, 149–62
severe combined immune deficiency (SCID),
 49–50, 61
sex
 determination of, 36–37
 X and Y chromosomes, 42–43
sexual intercourse, and spread of disease,
 92–93
 HBV, 111
 HCV, 111–12
 HIV, 222, 225–26
sexually transmitted diseases (STDs), 102,
 137
sheep, 202–3
 rendered into feed, 206
Shilts, Randy, 217–18, 235
shrimp, crazy, 129
silk industry, 69, 82
simian immunodeficiency virus, 218
simian virus 5, 109
sinuses, microbes in, 71
skin bacteria, 20
slaughter methods, and BSE, 210–11
smallpox, and demyelinating diseases, 108
Smith, Ralph, 197–98
Sontag, Susan, 78
sooty mangabey (*Cerocebus atys*), 219
Soviet Union, former, infectious disease
 increase in, 137
Spanish flu (of 1918), 238–40, 241
sperm, human, genetic material of, 47
spirochetes, 5–6
squid, quorum sensing in, 22–23
Staphylococcus aureus, 10, 12, 72, 86, 98–99,
 102, 139
 antibiotic-resistant, 99–101, 136
 in hospitals, 94
Staphylococcus epidermidis, 10
Stewart, William H., 85
stomach, digestive mechanism of, 114
stomach ulcers, 114

strep throat, and OCD, 125
Streptococcus mitis, 10
Streptococcus mutans, 10
Streptococcus pneumoniae, 10, 72, 92
Streptococcus pyogenes, 10
Streptococcus viridans, 10
streptomycin, 78
stroke, 115
sugar molecules, chirality of, 68–69
sugars, bacterial interactions with, 21
surgery, and risk of infection, 94
symbiosis, 30
symptoms, of same disease, differences in, 68
syphilis, 3–6, 14, 129, 137

Tamiflu, 251–52
tampons, 98–99
tartaric acid, 69
T-bone steaks, contamination of, 210
T-cell lymphoma, 86
T cells, 49, 232–33
testosterone, 36–37
T-helper lymphocytes (Th cells), 223–24
throat, microbes in, 71
thymidine, 39
tic disorders, 126
Tourette's disease, 126
tourism, and importation of infectious
 diseases, 138
toxic shock syndrome, 86, 98–99, 102
toxins, 194
Toxoplasma gondii, 123–24, 127
 cysts in brains, 123–24
trachea (windpipe), microbes in, 71
transmissible spongiform encephalopathies
 (TSEs), 129, 203–5
transplantation, liver, 111
travel, and resurgence of infectious disease,
 91–92, 138
trematodes (flatworms), spread of, 144–45
trench fever, 102
Treponema palidum, 5–6, 14, 129
Trichomonas tenax, 10, 14
tuberculosis (TB), 14, 71, 76–78, 102, 137,
 139
 antibiotic-resistant, 101
 antibiotics for, 77
 in U.S., 135, 136
 vaccine for, 77